Rapid Prototyping:
Theory and Practice

Rapid Prototyping:
Theory and Practice

Edited by

Ali Kamrani, Ph.D.
Industrial Engineering Department
University of Houston
Houston, TX, USA

and

Emad Abouel Nasr
Industrial Engineering Department
University of Houston
Houston, TX, USA

Library of Congress Cataloging-in-Publication Data

A C.I.P. Catalogue record for this book is available
from the Library of Congress.

ISBN-10: 0-387-23290-7 ISBN-10: 0-387-23291-5 (e-book)
ISBN-13: 9780387232904 ISBN-13: 9780387232911

Printed on acid-free paper.

© 2006 Springer Science+Business Media, Inc.
All rights reserved. This work may not be translated or copied in whole or in part without the written permission of the publisher (Springer Science+Business Media, Inc., 233 Spring Street, New York, NY 10013, USA), except for brief excerpts in connection with reviews or scholarly analysis. Use in connection with any form of information storage and retrieval, electronic adaptation, computer software, or by similar or dissimilar methodology now known or hereafter developed is forbidden.
The use in this publication of trade names, trademarks, service marks and similar terms, even if they are not identified as such, is not to be taken as an expression of opinion as to whether or not they are subject to proprietary rights.

Printed in the United States of America.

9 8 7 6 5 4 3 2 1   SPIN 11328148

springeronline.com

Preface

The current marketplace is undergoing an accelerated pace of change that challenges companies to innovate new techniques to rapidly respond to the ever-changing global environment. A country's economy is highly dependent on the development of new products that are innovative with shorter development time. Organizations now fail or succeed based upon their ability to respond quickly to changing customer demands and to utilize new innovative technologies. In this environment, the advantage goes to the firm that can offer greater varieties of new products with higher performance and greater overall appeal.

At the center of this environment is a new generation of customers. These customers have forced organizations to look for new methods and techniques to improve their business processes and speed up the product development cycle. As the direct result of this, the industry is required to apply new engineering philosophy such as ***Rapid Response to Manufacturing*** (RRM). RRM concept uses the knowledge of previously designed products in support of developing new products.

The RRM environment is developed by integrating technologies such as feature-based CAD modeling, knowledge-based engineering for integrated product and process design and direct manufacturing concepts. Product modeling within RRM requires advanced CAD technology to support comprehensive knowledge regarding the design and fabrication of a product. This knowledge-intensive environment utilizes knowledge-based technologies to provide a decision support utility throughout the design life cycle. Direct manufacturing uses rapid prototyping, tooling and manufacturing technologies to quickly verify the design and fabricate the part.

Rapid Prototyping (RP) is a technique for direct conversion of three dimensional CAD data into a physical prototype. RP allows for automatic construction of physical models and has been used to significantly reduce the time for the product development cycle and to improve the final quality of the designed product. Before the application of RP, computer numerically controlled (CNC) equipments were used to create prototypes either directly or indirectly using CAD data. CNC process consists of the removal of material in order to achieve the final shape of the part and it is in contrast to the RP operation since models are built by adding material layers after layers until the whole part is constructed. In RP process, thin-horizontal-cross sections are used to transform materials into physical prototypes. Steps in RP process are illustrated in Figure 1.

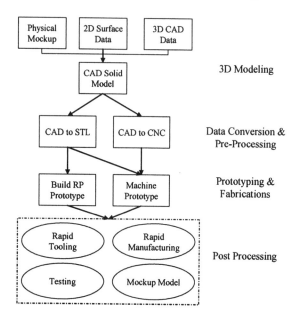

Figure 1. Generic RP Process

Depending on the quality of the final prototype, several iterated is possible until an acceptable model is built. In this process, CAD data is interpreted into the Stereolithugraphy data format. Stereolithugraphy or "stl" is the standard data format used by most RP machines. By using "stl", the surface of the solid is approximated using triangular facets with a normal vector pointing away from the surface in the solid. An example of triangulated surface using the stl format is illustrated in Figure 2.

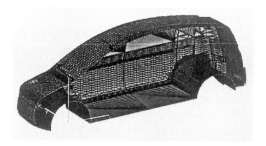

Figure 2. Triangulated surface

Since chordal deviation is used to approximate real mathematical surface, it is important to minimize this deviation to better approximate the real surface. This impact the size of the required triangles and it will also increase the processing time. Figure 3 illustrate the stages of product development using RP technologies.

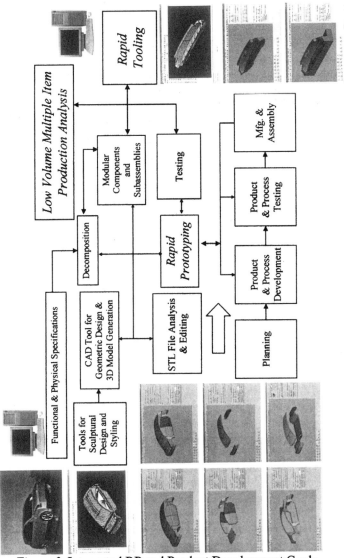

Figure 3. Integrated RP and Product Development Cycle

A wide range of technologies are developed to transform different materials into physical parts. For RP process, materials are categorized into liquid, solid and powdered. A process that has been widely used by many industries is the SLA or Stereolithography Apparatus RP Process. SLA is a liquid-based process. The material vat is part of the machine and is only removed if the liquid resin replaced. The SLA process uses the Ultraviolet laser beams to solidify the liquid polymer as it traces each layer. The part and the support are built simultaneously. The finished part is then manually removed, cleaned and finally post-cured using ultraviolet chamber. Another known process is the Fused Deposition Modeling or FDM. FDM is a solid-based process. The build material is melted inside an extrusion head where the temperature is contorted based on the type of the material used (ABS, wax, etc.). This semi-liquid material is then extruded and deposited layer-by-layer. The finished part is then manually removed and cleaned. Selective Laser Sintering (SLS) is a powder-based process. A CO_2 laser is raster scanned across the surface of the powder, melting and bonding the power together. In this process the part is built inside the powered material which can then be brushed off and reused. Sample parts are illustrated in Figure 4.

Figure 4. Sample RP parts

As Rapid Prototyping (RP) technology becomes more mature, it is beginning to lend itself to other applications such as rapid tooling and rapid manufacturing. Some traditional tool making methods are considering the use of RP technologies to directly or indirectly fabricate tools. The Indirect method of rapid tooling (RT) uses the RP pattern as mold. This is considered as a good alternative to the traditional mold making since it is more efficient and requires less lead-time. This approach is also less expensive and allows for quick validation of designs. In direct RT method, the RP process is used for direct fabrication of the tools. In summary, rapid tooling is described as a process which uses an RP model as:

1. A pattern to create mold quickly (e.g. sand casting),

2. Copy an RP form into a metal (e.g. investment casting), and
3. Uses the RP process directly to fabricate a limited volume of tools.

The next natural evolution of RP technology is Rapid Manufacturing (RM). RM is the automated fabrication of products directly from CAD digital data. RM methods are categorized into the following three categories:

1. *One-Time Use Parts*
   - Resemble the minimum required functionality of the final part.
   - Developed in batch-size of one and used for short duration.
   - High Cost.
2. *Individually Customized Parts*
   - Developed in batch-size of one for an indefinite period of time.
   - Durability of the parts is an issue and is based on the material used and its properties.
   - High Cost.
3. *Multiple Item Production Runs*
   - Methods for rapidly manufacture low volume production runs.
   - Not economically efficient to create mass quantities of identical parts using rapid manufacturing

In summary, various RP, RT and RM solutions are available and it is difficult for any organization to know which one is the most appropriate. It is recommended that companies compare and investigate advantages and disadvantages for all available methods and then select the one that is most suitable for their operational needs.

The purpose of this edited book is to provide a comprehensive collection of the latest research and technical work in the area of Rapid Prototyping, rapid tooling and rapid manufacturing. This book is developed to serve as a resource for researcher and practitioners. It can also be used as a text book for advanced graduate studies in product design, development and manufacturing. In chapter 1, Kridle gives an introduction to structure and properties of engineering materials, testing methods used to determine mechanical properties, and techniques that can be used to select materials for rapid prototyping. In chapter 2, Abouel Nasr and Kamrani will introduce a new methodology for feature extraction and information communication using IGES data. In chapter 3, Lim and Zein will introduce DICOM (Digital Imaging and Communications in Medicine) data format. DICOM is becoming a global information standard that is being used by virtually every medical profession that utilizes images within healthcare industry. Automatic feature recognition from CAD solid systems highly impacts the level of integration. Non contact-based reverse engineering is discussed in chapter 4 by Creehan and Binanda. In chapter 5, Desai and Binanda present the contacted-based reverse engineering process. In chapter 6 Kim and Nnaji present a discussion on virtual assembly. This chapter will discuss how assembly operation analysis can be embedded into a service-oriented collaborative assembly design environment and how the integrated process can help a designer to quickly select robust assembly design and process for rapid manufacturing. A new innovative RP process is presented by Frank in chapter 7. A description of how CNC milling can be used for rapid prototyping is presented in this chapter. The proposed methodology uses a layer–based approach for machining for automatic machining of common manufactured part geometries. Khoshnevis and Asiabanpour will introduce the SIS process in chapter 8. The Selective Inhibition of Sintering or SIS process is a new RP method that, like many other RP processes, builds parts in a layer-by-layer fabrication basis. The process works by joining powder particles through sintering in the part's body, and by sintering inhibition at the part boundary. In Chapter 9, Khoshnevis and Hwoang present Contour Crafting. CC is a mega scale fabrication technology based on Layered Manufacturing process (LM). This fabrication technique is capable of utilizing various types of materials to produce parts with high surface quality at high fabrication speed. Method for strategic justification of RP technologies is presented in chapter 10 by Narian and Sarkis. Wilson and Rosen give a discussion on a method for selection of a RM technology under the geometric uncertainty inherent to mass customization. This topic is presented in chapter 11. Specifically, they define the types of uncertainty inherent to RM, propose a method to account for this uncertainty in a selection process and propose a method to select a technology under uncertainty. In chapter 12, Gad El Mola and Parsaei present a methodology for selecting the best RP solution technique using Analytical Hierarchy Process (AHP).

Ali K. Kamrani, Ph.D.

To Sonia and Arshya

Acknowledgments

We would like to thank authors that participated in this project. We would also like to thank Mr. Steven Elliot and Ms. Rose Antonelli from Springer (US) publishing for giving us the opportunity to fulfill this project.

Contents

List of Figures	xxiii
List of Tables	xxix
Contributors	xxxi

Chapter 1: Material Properties and Characterization — 1
    Ghassan T. Kridli.

1.1. Structural Properties of Materials	2
1.1.1. Crystalline Structures	2
1.1.2. Non-crystalline (Amorphous) Structures	5
1.2. Engineering Material Classification	6
1.2.1. Metals	6
1.2.2. Ceramics and Glass	7
1.2.3. Polymers	7
1.2.4. Composite Materials	8
1.3. Mechanical Properties of Materials	9
1.3.1 Uniaxial Tension Test	9
1.3.1.1. Tensile Modules	10
1.3.1.2. Engineering Stress	11
1.3.1.3. Engineering Strain	13
1.3.1.4. Ductility	14
1.3.1.5. True Stress and True Strain	14
1.3.2. Toughness Tests	15
1.3.3. Hardness Tests	16
1.3.4. Flexure Tests	16
1.3.5. Creep Test	17
1.4. Polymers used in Rapid Prototyping	18
1.5. Material Selection	20

Part I. Direct and Indirect Data Input Formats

Chapter 2: IGES Standard Protocol for Feature Recognition
    CAD System   25
Emad Abouel Nasr and Ali Kamrani

2.1. History and Overview	26
2.2. Standard data format	27
2.2.1. Data transfer in CAD/CAM Systems	27
2.2.2. Initial Graphics Exchange Specifications (IGES)	29
2.2.2.1. Structure of IGES File	30
2.2.2.1.1. Start Section	31
2.2.2.1.2. Global section	31
2.2.2.1.3. Directory entry section (DE)	31
2.2.2.1.4. Parameter data section (PD)	32
2.2.2.1.5. Terminate section	32
2.3. The Feature extraction Methodology	33
2.3.1. Conversion of CAD data files into Object Oriented Data structure	34
2.3.1.1. Basic IGES entities	34
2.3.2. The Overall object-oriented data structure of the proposed methodology	37
2.3.2.1. Geometry and topology of B-rep	41
2.3.2.1.1. Classification of Edges	41
2.3.2.1.2. Classification of loops	42
2.3.2.2. Definition of the Data Fields of the Proposed Data structure	44
2.3.2.3. Algorithms for Extracting Geometric Entities from CAD File	44
2.3.2.3.1. Algorithm for extracting entries from directory and parameter sections	46
2.3.3.3.2. Algorithm for extracting the basic entities of the designed part	47
2.3.3.4 Extracting Form Features from CAD Files	49
2.3.3.4.1. An Example for identifying the concave edge/faces	51
2.3.3.4.2. Algorithm for determining the concavity of the edge	51
2.3.3.4.3. Algorithms for feature extraction (Production rules)	52
2.4. An Illustrative Example	54
2.5. Summary	60

Chapter 3: The Digital Imaging and Communications in Medicine (DICOM): Description, Structure and Applications   63
Jinho (Gino) Lim and Rashad Zein

3.1. Introduction	64
3.1.1. History	64
3.1.2. The scope of current DICOM Standards	67
3.2. DICOM structure	67
3.2.1. Entity- Relationship models	68
3.2.2. DICOM components	70
3.2.3. DICOM Format	72
3.3. Current applications	75
3.4. Use of DICOM in Radiation Treatment Planning	78
3.5. Potential use of DICOM as a tool in various Industries	83
3.6. Summary	84

Chapter 4: Reverse Engineering: A Review & Evaluation of Non-Contact Based Systems   87
Kevin D. Creehan and Bopaya Bidanda

4.1. Introduction	88
4.2. Non-contact reverse engineering Techniques	89
4.2.1. Reverse Engineering Taxonomy	89
4.2.2. Active Technique – Laser Scanning	91
4.2.3. Passive Technique – Three-Dimensional Photogrammetry	94
4.2.4. Medical Imaging	96
4.2.4.1. Magnetic Resonance Imaging	96
4.2.4.2. Computed Tomography	98
4.2.4.3. Ultrasound Scanning	99
4.2.4.4. Medical Image Data File	99
4.2.4.5. Three-Dimensional Reconstruction	101
4.3. Applications	102
4.4. Relationship to rapid prototyping	104

Chapter 5: Reverse Engineering: A Review & Evaluation of Contact Based Systems   107
Salil Desai and Bopaya Bidanda

5.1. Introduction	108
5.1.1. Need for reverse engineering	108
5.2. Contact Based Reverse Engineering Systems	109
5.3. Coordinate Measuring Machine (CMM)	110
5.3.1. Types of CMM Configurations	111

5.3.1.1. Bridge Type	111
5.3.1.1.1. Applications	111
5.3.1.2. Gantry type	112
5.3.1.2.1. Applications	113
5.3.1.3. Cantilever Type	113
5.3.1.3.1. Applications	114
5.3.1.4. Horizontal Arm Type	114
5.3.1.4.1. Applications	114
5.3.1.5. Articulated Arm Type	115
5.3.1.5.1. Applications	115
5.3.2. Specifications of Coordinate Measuring machines	116
5.3.2.1. Control types	116
5.3.2.2. Mounting options	116
5.4. CMM Measurement process	117
5.4.1. Data Collection Procedure for CMM	118
5.4.2. Digitization from the Surface	119
5.4.3. Preprocessing of The Point Clouds	119
5.4.3.1. Point processing as applied to a Knee Joint	120
5.4.4. Surface fitting	121
5.5. Performance Parameters of CMMs	121
5.5.1. Scanning speed	121
5.5.2. CMM probe accuracy	122
5.5.3. CMM Rigid Body Errors	123
5.5.4. CMM Structural Deformations	124
5.6. Integration of CMM data into other design and manufacturing System Software	124
5.7. Recent advances in CMM technology	125
5.7.1. Reverse Engineering method based on Haptic Volume Removal	125
5.7.2. Nano CMM	127

Part II. Methods and Techniques

Chapter 6: VIRTUAL ASSEMBLY ANALYSIS ENHANCING RAPID PROTOTYPING IN COLLABORATIVE PRODUCT DEVELOPMENT 133
Kyoung-Yun Kim and Bart O. Nnaji

6.1. History and Overview	134
6.2. Modern Design and product development	136
6.2.1 Rapid product development	136
6.2.2 Internet-enabled collaboration	137
6.2.3 Inevitable impact on assembly and joining Operations	138

6.3. Collaborative Virtual Prototyping and simulation	140
6.4. Service-oriented Collaborative Virtual Prototyping and Simulation	140
6.5. Virtual Assembly analysis	143
6.5.1 Service-Oriented VAA architecture and Components	144
6.5.2 VAA tool	145
6.5.2.1. Assembly design formalism and assembly design model generation	146
6.5.3. Assembly analysis model (AsAM) generation	147
6.5.4. Pegasus Service Manager	149
6.5.5. e-Design Brokers	150
6.5.6. Service Providers	150
6.6. Implementations	151
6.7. Contributions	159
6.8. Conclusions	160

Chapter 7: Subtractive Rapid Prototyping: Creating a Completely Automated Process for Rapid Machining   165
    Matthew C. Frank

7.1. Background	166
7.2. Related Work	168
7.3. Assumptions	169
7.4. Overview of the CNC-RP Process	170
7.5. Approach to setup Planning	174
7.6. Approach to Tool Selection	178
7.7. Challenges with Rapid Fixturing	180
7.8. General System model	184
7.9. Example parts using CNC-RP	185
7.10. Economics of CNC-RP	189
7.11. Limitations and Future Work	192
7.12. Summary	194

Chapter 8: SELECTIVE INHIBITION OF SINTERING   197
    Behrokh Khoshnevis and Bahram Asiabanpour

8.1. Introduction	197
8.2. The sis process materials	202
8.3. The sis process machine path generation	204
8.3.1. Step 1: slicing algorithm	204
8.3.2. Step 2: machine path generation	206
8.4. The SIS process optimization	210
8.5. Physical part fabrication	216
8.6. Powder waste Reduction	217
8.7. Vision for future Research	218

Chapter 9: Contour Crafting: A Mega Scale Fabrication Technology 221
Behrokh Khoshnevis and Dooil Hwang

9.1. Introduction	222
9.1.1. CC Process error analysis	223
9.1.2. CC Applications	226
9.1.2.1. Ceramic part Fabrications	226
9.1.2.2. CC machine structure for Ceramics Processing	227
9.1.2.3. Preparing ceramic Paste	227
9.1.2.4. Prefabricated Ceramic Parts	228
9.1.2.5. Fabrication of pre-functional piezoelectric lead zirconate titanate (PZT) ceramic components and thermo-plastic parts	228
9.1.2.6. Design considerations	230
9.1.2.7. Experiments	232
9.1.3. Construction Automation	234
9.1.3.1. Challenges faced by Construction industry Today	235
9.1.3.2. The current state of automation in construction sites	236
9.1.3.3. Barriers for construction automation	237
9.1.3.4. The needs for an innovative construction process	238
9.1.3.5. CC concrete formwork design	239
9.1.3.6. Physical properties of CC formwork material	241
9.1.3.7. CC nozzle geometry	242
9.1.3.8. Fabrication of vertical concrete formwork	244
9.1.3.9. Placing fresh concrete	245
9.1.3.10. Results	247
9.1.3.11. Extraterrestrial construction	248
9.1.4. Conclusion	249

Part III. Economic Analysis

Chapter 10: Strategic Justification of Rapid Prototyping Systems 253
Rakesh Narain and Joseph Sarkis

10.1. Introduction	254
10.2. Investment justification factors	256
10.2.1. Strategic and Operational Benefits	256
10.2.2. Cost	257
10.2.3. Systems Characteristics and Factors	258
10.2.3.1. Inter- and Intra-Firm Adaptability	258
10.2.3.2. Platform Neutrality and Interoperability	258
10.2.3.3. Scalability	258
10.2.3.4. Reliability	258

10.2.3.5. Ease of Use	259
10.2.3.6. Customer Support	259
10.3. Issues and Models for Evaluation and Justification of RP	259
10.3.1. The Analytical Network Process (ANP) Methodology	260
10.3.1.1. Step 1: Setting up the Network Decision Hierarchy	261
10.3.1.1.1. The Planning Horizon	263
10.3.1.2. Step 2: Pairwise Comparisons	263
10.3.1.3. Step 3: Calculate Relative Importance Weights	264
10.3.1.4. Step 4: Form a Supermatrix	265
10.3.1.5. Step 5: Arrive at a Converged set of Weights	265
10.4. Summary and Discussion	267

Chapter 11: SELECTION OF RAPID MANUFACTURING TECHNOLOGIES UNDER EPISTEMIC UNCERTAINTY — 271
Jamal O. Wilson and David Rosen

11.1. Introduction	272
11.2. Uncertainty and its representations	273
11.3. Selection for rapid manufacturing	277
11.4. Illustrative example: direct production of caster wheels	281
11.5. Illustrative example: direct production of hearing aid shells	286
11.6. Closure	289

Chapter 12: Economic Analysis of Rapid Prototyping Systems — 293
Khaled Gad El Mola, Hamid Parsaei, and Herman Leep

12.1. Introduction	294
12.2. Recent Literature on Justification Techniques	295
12.3. Analytic Hierarchy Process and Expert Choice	299
12.4. Analytical Model to Justify Advanced Manufacturing Technologies	301
12.4.1. Identify the competitive criteria and their measures	302
12.4.2. Structuring the hierarchy	302
12.4.3. Determine the overall weight for each alternative and select the alternative that has the highest weight	303
12.5. An Illustrated Example	305
12.6. Summary	315

Index — 319

List of Figures

Figure 1-1. Schematic of common crystal structure — 3
Figure 1-2. Thermal expansion behavior of crystalline materials — 4
Figure 1-3. Examples of crystal imperfections — 4
Figure 1-4. Structures for Crystalline and Non-Crystalline Materials — 5
Figure 1-5. Thermal expansion behavior of an amorphous material — 6
Figure 1-6. Polymer Structures — 8
Figure 1-7. Uniaxial tensile test specimen shape — 10
Figure 1-8. Load-displacement behavior in metals and polymer — 11
Figure 1-9. Normal load, Pn, and shear loads, Ps — 12
Figure 1-10. Schematic showing the original and the deformed sizes of the gage section of a uniformly stretched tensile test specimen — 13
Figure 1-11. Schematic of the specimen setup in Charpy impact test — 15
Figure 1-12. Schematic of test specimen, loading, and support in flexure tests — 17
Figure 2-1. Solid Modeling technology evolutions — 28
Figure 2-2. Translation using a Neutral File — 29
Figure 2-3. IGES translators — 30
Figure 2-4. IGES file structure — 31
Figure 2-5. Structure of Directory Section — 32
Figure 2-6. Structure of Parameter Data Section — 33
Figure 2-7. Structure of the proposed methodology — 36
Figure 2-8. Flowchart of extraction and classification of features — 37
Figure 2-9. Hierarchy of classes and attributes of the designed object — 38
Figure 2-10. Simple and Compound Features — 39
Figure 2-11. Convex and Concave Features and Edges — 40
Figure 2-12. Classification of Interior and Exterior Form Features — 40
Figure 2-13. Classifications of Convex Features — 41
Figure 2-14. The surface normal vectors — 42
Figure 2-15. Classification of Edges — 43
Figure 2-16. Classification of Loops — 43
Figure 2-17. The Direction of Edge — 49

Figure 2-18. A concave Edge Example — 50
Figure 2-19. STEP THROUGH — 52
Figure 2-20. STEP THROUGH ROUND CORNER — 53
Figure 2-21. Illustrative Example — 54
Figure 3-1. A CT image — 64
Figure 3-2. The DICOM communications protocol model. — 66
Figure 3-3. E-R model — 69
Figure 3-4. DICOM components. — 70
Figure 3-5. A DICOM image file. — 72
Figure 3-6. A DICOM header as displayed by MRIcro. — 73
Figure 3-7. A sample of a Transfer Syntax UID. — 74
Figure 3-8. The way DICOM defines the interface — 75
Figure 3-9. DICOM images in the dental field. — 78
Figure 3-10. A DICOM image displayed in DICOMWORKS. — 80
Figure 3-11. DICOM image information in text format. — 80
Figure 3-12. Displaying a treatment plan using CERR. — 82
Figure 3-13. Dose volume histogram (DVH). — 82
Figure 3-14. DICOM applications in rapid prototyping. — 84
Figure 4-1. Reverse engineering taxonomy2. — 91
Figure 4-2. The triangle formed between the laser, the scanned part, and the sensors. — 92
Figure 4-3. Three-dimensional laser scanners. — 93
Figure 4-4. The triangles formed between the multiple image Perspectives — 94
Figure 4-5. Samsung SCC-B2305 CCD camera — 96
Figure 4-6. General Electric 3.0 Tesla Signa VH MRI scanner. — 97
Figure 4-7. General Electric Sytec 1800i CT scanner. — 99
Figure 4-8. Relationship between reverse engineering and rapid prototyping. — 104
Figure 5-1. Examples of Bridge type CMMs. — 112
Figure 5-2. Examples of Gantry type CMMs. — 113
Figure 5-3. Examples of Cantilever type CMMs. — 113
Figure 5-4. Dual Arm Type Horizontal CMM. — 114
Figure 5-5. Articulated Arm Type CMM. — 115
Figure 5-6. Schematic Diagram for a CMM. — 117
Figure 5-7. Scanning Probe path. — 118
Figure 5-8. Re-sampling process for mesh generation. — 118
Figure 5-9. CMM scanning process. — 119
Figure 5-10. Point cloud before and after preprocessing. — 120
Figure 5-11. Measurement accuracy during slow and fast speed scanning — 122
Figure 5-12. Haptic volume sculpting based reverse engineering. — 126
Figure 5-13. Construction of a Nano CMM. — 128
Figure 5-14. Prototype of the Nano CMM. — 129
Figure 5-15. Construction and prototype of a nanoprobe. — 129

Figure 6-1. Rapid prototyping in product development.	137
Figure 6-2. Service triangular relationship.	142
Figure 6-3. Collaborative virtual prototype model generation.	142
Figure 6-4. Virtual assembly analysis.	143
Figure 6-5. Service-oriented VAA architecture.	145
Figure 6-6. Service transactions in VAA.	153
Figure 6-7. Assembly models for VAA.	154
Figure 6-8. Pegasus Service Manager.	155
Figure 6-9. VAA service provider.	155
Figure 6-10. Equivalent stress and deformation obtained from VAA Service	157
Figure 6-11. Aluminum concept car and body frames (Buchholz 43).	157
Figure 6-12. VAA for a welded extruded frame.	158
Figure 6-13. VAA for a hinge with three rivets.	159
Figure 7-1. Free-form surface being machined from two orientations.	171
Figure 7-2. Setup for CNC-RP.	172
Figure 7-3. Process steps for CNC-RP.	173
Figure 7-4. A sample part.	174
Figure 7-5. Model with sample cross section used for visibility mapping.	175
Figure 7-6. Visible ranges of one segment of a polygonal chain.	176
Figure 7-7. Layer-based toolpath boundaries.	177
Figure 7-8. Distance to the deepest visible segment at one orientation.	177
Figure 7-9. Tool length requirement.	175
Figure 7-10. Tool diameter requirement.	179
Figure 7-11. Cutter contact area for flat- and ball-end mill.	179
Figure 7-12. Comparison of traditional vs. feature-free fixturing.	181
Figure 7-13. Setup for Rapid Machining.	183
Figure 7-14. CNC-RP System Model.	184
Figure 7-15. Jack model.	185
Figure 7-16. Bone model.	186
Figure 7-17. Visibility orientations for the femur.	187
Figure 7-18. Finished prismatic model.	187
Figure 7-19. SLA model.	188
Figure 7-20. Example models.	190
Figure 7-21. Example part with non-orthogonal feature.	191
Figure 7-22. Parts with no feasible axis.	193
Figure 8-1. Stages of the SIS Process.	198
Figure 8-2. Extraction of the Fabricated Part.	198
Figure 8-3. The Selective Inhibition Sintering Alpha Machine.	200
Figure 8-4. Selective Inhibition Sintering Process Steps.	201
Figure 8-5. Water Evaporation in the Selected Area and Salt Wash Off to Extract the Part.	202
Figure 8-6. Salt Wash Off and Part Extraction.	203
Figure 8-7. Steps for a sample part production using KI solution	203

Figure 8-8. The Two Steps of Slicing and Machine Path Generation 204
Figure 8-9. Slicing Algorithm Steps. 205
Figure 8-10. Machine Path Generation Steps. 207
Figure 8-11. Hatch Path Generation Steps. 208
Figure 8-12. Visualization CAD model and the machine path (left) 210
Figure 8-13. IDEF0 Hierarchy for the SIS Process. 210
Figure 8-14. Roadmap for the SIS Process Properties Optimization. 211
Figure 8-15. Base Part. 212
Figure 8-16. Part Breaking Mechanism. 212
Figure 8-17. Shift_X. 213
Figure 8-18. Part Breakdown for Surface Quality Rating. 214
Figure 8.19. General Form of the Component Desirability Function. 215
Figure 8- 20. 2.5D Parts Fabricated by the SIS Process. 216
Figure 8- 21. 3D Parts Fabricated by the SIS Machine. 216
Figure 8-22. Current SIS Process Limitations: Waste Material. 217
Figure 8-23. Moveable Fingers Mask System. 217
Figure 8-24. The New Mask System Design (top) and Prototype (bottom) 218
Figure 9-1. Schematic of CC extrusion and filling process 222
Figure 9-2. Schematic of trowels and extrusion assembly 222
Figure 9-3. Cross-section comparisons of layered boundaries 223
Figure 9-4. Geometric description of local layered process error 224
Figure 9-5. Contour Crafting nozzle and some of 2.5D geometries and 3D parts 226
Figure 9-6. CC system configurations for ceramic part fabrications 227
Figure 9-7. Demonstrations of CC constructability 228
Figure 9-8. Advanced ceramic structures. 229
Figure 9-9. An adapted CC system for fabricating pre-functional ceramic and thermoplastic parts. 230
Figure 9-10. Material feeding system with adjustable length (L) between rollers and heating barrel 213
Figure 9-11. Schematic of CC working platform to prevent thermal contraction 231
Figure 9-12. Schematic of die shape designs and sectional view of layers 232
Figure 9-13. Showing customized nozzle shape and its actual fabrication 232
Figure 9-14. Fabricated 2.5D geometries and two advanced structure shapes 233.
Figure 9-15. Pre-functional advance ceramic (Spiral PZT actuator) 234
Figure 9-16. Schematic of the automatic construction of a residential building. 235
Figure 9-17. A single task robot, Surf Robo.236
Figure 9-18. The first automated construction system applied steel concrete structure. 237
Figure 9-19. Basic components of a typical wall formwork 240
Figure 9-20. Closer sections of wall form. 241

Figure 9-21. Excessive friction force causes some voids
    on extruded flow. 243
Figure 9-22. Specialized CC nozzle assemblies. 243
Figure 9-23. Extrusion flow control by new CC trowels. 244
Figure 9-24. A concrete wall form fabricated by the CC machine. 245
Figure 9-25. CC concrete placing procedures. 246
Figure 9-26. Placing concrete in layer by layer without using form ties. 247
Figure 9-27. A concrete wall made by CC machine. 248
Figure 11-1. Certainty Equivalent Determination. 275
Figure 11-2. Hurwicz Selection Criteria 276
Figure 11-3. The Word Formulation for Selection for Rapid
    Manufacturing 277
Figure 11-3. Summary of Steps for Selection for Rapid Manufacturing. 277
Figure 11-5. Model of steel caster wheel 282
Figure 11-6. Hearing Aid Shell. 286
Figure 12-1. Hierarchical Structure for Justification of Advanced
    Manufacturing Technologies 304
Figure 12-2. AHP Model for Selecting the Best AMT Alternative 306
Figure 12-3. A Graphical Pairwise Comparison of Economic
    and Non-Economic Benefits with Respect to
    the Overall Objective 307
Figure 12-4. A Numerical Pairwise Comparison of
    Non-Economic Criterion 308
Figure 12-5. Relative Weights with Respect to Non-Economic Benefits 308
Figure 12-6. A Verbal Pairwise Comparison of Economic Criterion 309
Figure 12-7. Relative Weights of Alternative with Respect to
    Product Change Response 311
Figure 12-8. Relative Weights of Alternatives with Respect to
    Design Time 311
Figure 12-9. Relative Weights of Alternatives with Respect to
    Cycle Time 312
Figure 12-10. Relative Weights of Alternatives with
    Respect to Delivery Reliability 312
Figure 12-11. Relative Weights of Alternatives with
    Respect to ROI 313
Figure 12-12. Relative Weights of Alternatives with Respect
    to Start-Up Costs 313
Figure 12-13. Relative Weights of Alternatives with respect
    to Tooling Costs 314

List of Tables

Table 1-1. Selected properties of relevant polymers	22
Table 2-1. Classification of Loops	44
Table 2-2. Definitions of classes and attributes	45
Table 2-3. Extraction of Vertices	55
Table 2-4. Extraction of Edges	55
Table 2-5. Extraction of Loops	57
Table 2-6. Extraction of Faces	58
Table 2-7. Extraction of Features	59
Table 2-8. Machining Information	60
Table 6-1. AsAM for welding	148
Table 6-2. AsAM for riveting	148
Table 7-1. Build and Post Process Times	188
Table 7-2. Comparison of Total Processing Time	188
Table 8-1. Part surface quality rating system	214
Table 9-1. Results of testing compressive strength	242
Table 10.1. Example Pairwise Comparison Matrix and resulting Relative Importance Weight for Strategic Metrics	264
Table 10-2. Initial Supermatrix	266
Table 11-1. Caster wheel dimensions	282
Table 11-2. Attribute Ratings	284
Table 11-3. Alternative Merit Function Values for Scenario 1 and 2	285
Table 11-4. Hurwicz evaluation parameters	285
Table 11-5. Hearing Aid Shell Dimensions	287
Table 11-6. Attribute Ratings	288
Table 11-7. Merit Function Values and Hurwitz factors	289
Table 12-1. Summary of Available Justification Approaches	300

Contributors

Ghassan T. Kridli, PhD
Industrial and Manufacturing Systems Engineering Department, University of Michigan Dearborn, 4901 Evergreen Rd., Dearborn, MI, 48128.

Emad Abouel Nasr
Industrial Engineering Department, Faculty of Engineering, University of Houston, 213 Engineering Building 2, Houston, TX 77204-4008.

Ali K. Kamrani, Ph.D.
Industrial Engineering Department, Faculty of Engineering, University of Houston, 213 Engineering Building 2, Houston, TX 77204-4008.

Jinho (Gino) Lim, Ph.D.
Industrial Engineering Department, University of Houston, 213 Engineering Building 2, Houston, TX 77204-4008.

Rashad Zein
Industrial Engineering Department, University of Houston, 213 Engineering Building 2, Houston, TX 77204-4008.

Kevin D. Creehan, Ph.D.
Center for High Performance Manufacturing, Department of Industrial and Systems Engineering, Virginia Tech, Blacksburg, VA 24061.

Salil Desai, Ph.D.
Department of Industrial & Systems Engineering, NC A&T State University, Greensboro, NC 27411.

Bopaya Bidanda, Ph.D.
Department of Industrial Engineering, University of Pittsburgh, Pittsburgh, PA 15261

Kyoung-Yun Kim, Ph. D.
Department of Industrial and Manufacturing Engineering, Wayne State University, 4815 Fourth St., Detroit, MI 48201.

Bart O. Nnaji, Ph.D.
US NSF Center for e-Design and Department of Industrial Engineering, University of Pittsburgh, Pittsburgh, PA 1526.

Matthew C. Frank, Ph.D.
Department of Industrial and Manufacturing Systems Engineering, Iowa State University, Ames, IA 50011.

Bahram Asiabanpour , Ph.D.
Texas State University-San Marcos, TX 78666.

Behrokh Khoshnevis, Ph.D.
Professor, Department of Industrial and Systems Engineering, University of Southern California, USA

Dooil Hwang, Ph.D.
Research Associate, Information Sciences Institute, University of Southern California, US.

Rakesh Narain, Ph.D.
Department of Mechanical Engineering, Motilal Nehru National Institute of Technology, Allahabad-211004, India.

Joseph Sarkis, Ph.D.
Clark University, Graduate School of Management, 950 Main Street, Worcester, Massachusetts, 01610.

Jamal O. Wilson, Ph.D.
Systems Realizations Laboratory, The G.W. Woodruff School of Mechanical Engineering, Georgia Institute of Technology, Atlanta, GA 30332-0405.

David Rosen, Ph.D.
Systems Realizations Laboratory, The G.W. Woodruff School of Mechanical Engineering, Georgia Institute of Technology, Atlanta, GA 30332-0405.

Khaled M. Gad El Mola, Ph.D.
Department of Industrial Engineering, University of Houston, Houston, TX 77204.

Hamid R. Parsaei, Ph.D.
Professor and Chairman, Department of Industrial Engineering, University of Houston, Houston, TX 77204.

Herman R. Leep, Ph.D.
Professor, Department of Industrial Engineering, University of Louisville, Louisville, KY 40292.

Chapter 1

Material Properties and Characterization

Ghassan T. Kridli

Industrial and Manufacturing Systems Engineering Department, University of Michigan Dearborn, 4901 Evergreen Rd., Dearborn, MI, 48128

Abstract:

As the name indicates, rapid prototyping (RP) has traditionally been used to provide a physical representation of a product in a relatively short time. RP is performed by either material removal or material addition. In material-removal type RP processes, the part is produce by machining it is out of a block of material; mainly using computer numeric controlled (CNC) machining centers. In material-addition type RP, the prototype is made by adding layers of materials using one of the available RP technologies.
Earlier prototyping materials and technologies were used to provide product designers with the ability to visualize the product, but with limited ability to assess the functional performance of the product. Nonetheless, prototyped parts also need to allow for design validation (assessment of the mechanical and physical behaviors); which indicates that the prototyping material should have the same characteristics as the production material. This was only available in limited situations where the prototyped parts were made using removal processes, casting processes, or metal spray deposition. However, recent advances in rapid prototyping technologies have

allowed the use of production type polymers that can be used to assess the functional behavior of these materials.

One of the shortcomings of testing prototyped products made of production type materials is that the material structure and the mechanical response of the prototyped part may not match those resulting from conventional processing (forming, molding, etc.) that is used to fabricate the actual product. This is caused by differences in processing conditions between RP and conventional processing. For example, if metal spray deposition is used for rapid prototyping purposes, the microstructure and level of porosity in the prototyped part are likely to be different from those of a cast or stamped product of the same size and shape.

Therefore, the goal of this chapter is to provide an introduction to structure and properties of engineering materials, testing methods used to determine mechanical properties, and techniques that can be used to select materials for material-addition type rapid prototyping.

Key words:
Mechanical Properties, Mechanical Testing, Material Selection, Polymers

1.1 Structural Properties of Materials

The structure of materials affects their properties and service behavior. Based on their structure, materials can be classified as either crystalline or non-crystalline (or amorphous)[1]. **Crystalline structures** are organized structures in which atoms and molecules of solids arrange themselves in a regular and repeating manner that is called *lattice*. On the other hand, **amorphous structures** have some level of local order relative to their neighbors, but globally, they do not have an ordered structure like crystalline materials. Another difference between the two types of materials is related to their different thermal expansion behavior; this will be explained in more detail in sections 1.2.1 and 1.2.2.

1.1.1 Crystalline Structures

The lattice structure in a crystalline material is made up of a repeating order of atoms that is known as the *unit cell*. Common types of unit cells that are observed in metals include body-centered cubic (BCC), face-centered cubic (FCC), and hexagonal close packed (HCP). Figure 1-1 shows a schematic of the atom arrangement in each of these three aforementioned types of structures. For clarification purposes, atoms/molecules in Figure 1-1 that are at the vertices of the unit cell are shaded with a solid pattern, while those atoms in the middle of faces of the unit cell (or in the middle of the unit cell as in BCC) are shaded with a pattern.

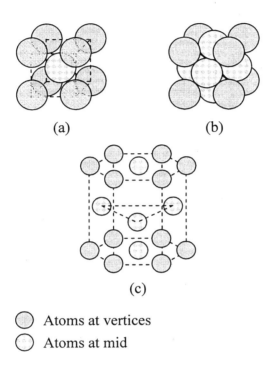

Figure 1-1. Schematic of common crystal structures (a) BCC, (b) FCC, and (c) HCP

Upon heating, a crystalline material undergoes uniform **thermal expansion** (change in volume per unit weight as temperature increases) until it reaches its **melting temperature**, T_m, at which the material turns into a liquid. The heat required to transform the material from a solid at T_m to a liquid also at T_m is called the **heat of fusion**. After the material melts, it begins to expand at a higher rate, as can be seen in Figure 1-2.

Three types of defects or imperfections are typically observed in crystalline structure.

1. Point defects such as **vacancy** type defects, where an atom is missing at one of the locations in the unit cell.

2. Line defects such as edge **dislocations**, where an additional half plane of atoms is present in the lattice.
3. Surface defect such as **grain boundaries** separating crystals (grains).

Schematics of the examples given for each type of defect are presented in Figure 1-3. These defects along with the material structure play an important role in determining the material properties.

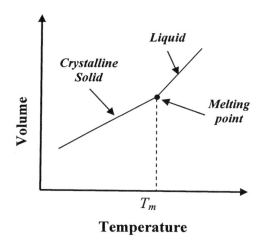

Figure 1-2. *Thermal expansion behavior of crystalline materials*

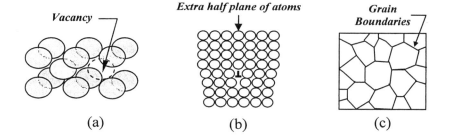

Figure 1-3. Examples of crystal imperfections (a) vacancy, (b) edge dislocation, and (c) grain boundaries.

1.1.2 Non-crystalline (Amorphous) Structures

Amorphous structures are generally observed in glass and plastic materials, even though some plastics also have a **semi-crystalline structure**, which is a combination of crystalline and amorphous structures. As can be seen from Figure 1-4, the atoms/molecules are less closely packed in an amorphous structure compared with a crystalline structure, thus, for a given material; the density of an amorphous structure is lower than that of a crystalline structure.

As an amorphous material is heated, it expands at a constant rate until it reaches a point beyond which the solid starts to expand at a higher rate. This point is known as the **glass transition temperature**, T_g. The material continues to expand at this higher rate even as it is transforming into a liquid. The solid material between T_g and T_m is a transition/nonequilibrium phase between a true liquid state and an amorphous solid state, and is referred to as **supercooled liquid**[1]. Unlike crystalline structures, there is no sudden change in thermal expansion coefficient at the melting point of an amorphous material as can bee seen in Figure 1-5. It is should be noted that T_g depends on cooling rate; therefore, a range is specified for T_g in polymers[2].

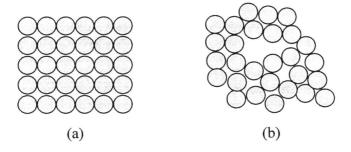

(a) (b)

Figure 1-4. Structures for (a) Crystalline Materials and (b) Non-Crystalline Materials

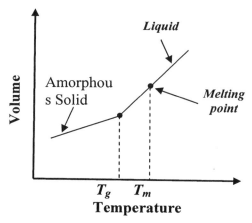

Figure 1-5. Thermal expansion behavior of an amorphous material

1.2 Engineering Material Classification

Engineering materials are materials that carry loads and/or provide capabilities to meet functional requirements, such as electrical and thermal conductivities[2]. Engineering materials are typically classified into four classes: metals, ceramics and glasses, polymers, and composite materials. A general description of each class of engineering materials is provided in this section. Materials of all types can be used for rapid prototyping purposes.

1.2.1 Metals

Metals are the most commonly used type of engineering materials. Metals are extracted from ores either directly or indirectly[1]. Metals have crystalline structures that are most commonly of the types BCC (such as in iron), FCC (such as in aluminum), and HCP (such as in magnesium). Metals possess several attractive properties including their **yield strength** (maximum stress that can be applied without causing plastic, or permanent, deformation), **ultimate tensile strength** (maximum stress that can be applied without causing necking), **stiffness** (resistance to deformation or deflection), electrical and thermal conductivities, **toughness** (ability to absorb energy without fracture), **hardness** (resistance to deformation by indentation) as well as many others.

Metals are available in three forms (i) cast metals (ii) wrought metals, and (iii) powder metals. Cast metals are primarily fabricated into useful products through melting and casting into shapes. Wrought metals are processed to have a desired structure through multiple hot or hot and cold rolling operations starting from a cast slab. They are used in the manufacture of products through deformation processes. Powder metals are used for powder metallurgy processes. Metals are also classified as ferrous and non-ferrous metals. Ferrous metals are iron-based metals such as steels and cast irons, and non-ferrous metals include all other metals.

The properties of pure metals can be enhanced through the production of alloys. An **alloy** is produced through the addition of other elements to the metal; the alloying element(s) need not be metallic. For example, steel is produced by adding carbon to iron; the amount of carbon in steel directly influences its strength.

1.2.2 Ceramics and Glass

Ceramics can have either a crystalline structure or an amorphous structure. They are widely available in nature and are inorganic materials; however, there crystalline structure is different from that observed in metals so that different sized atoms are accommodated in the crystal[3]. Ceramics are compounds of a metal or semi-metal and at least one more non-metallic material. According to Groover[1], ceramics are classified into three types:
1. Traditional ceramics that are base on silicates; such as clay and cement products.
2. New ceramics that are not based on silicates; such as alumina or aluminum oxide (Al_2O_3) and cubic boron nitride (CBN).
3. Glasses are silica (SiO_2) based and have an amorphous structure.

1.2.3 Polymers

A polymer is made of multiple repeating molecular structures. The name polymer comes from two Greek words *poly* and *meres* meaning "many parts"[4]. Polymers have an amorphous or a **semi-crystalline** structure, which is a combination of crystalline and amorphous structures. Polymers have lower density, strength and stiffness than metals. They also have very low electrical conductivities, which makes them suitable for use as insulation material. Additional information on the types of polymers that are used in rapid prototyping processes is provided in Section 4. Polymers are classified into three groups:

1. **Thermoplastics**: have a linear or branched structure and can be re-softened by heating and then reshaped.
2. **Thermosets**: have cross-linked structures that develop upon **curing** (controlled heating to promote cross linking). Once cured, they cannot be reshaped.
3. **Elastomers**: also have a cross-linked structure. They possess elastic behavior similar to that of natural rubber.

A schematic showing linear and cross-linked type polymers can be seen in Figure 1-6.

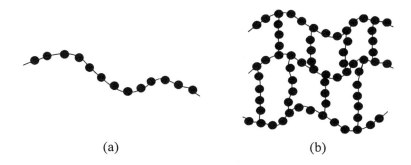

(a) (b)

Figure 1-6. [Polymer Structures (a) Liner and (b) Cross-Linked (each circle represents a mer)

1.2.4 Composite Materials

The term composite material refers to a solid material made by combining two or more different materials, by mechanical and metallurgical means, to produce a material with enhanced properties compared to each of its constituents. Composite materials consist of a **matrix material** and the reinforcing material that is imbedded in the matrix material. In addition to holding the reinforcing material, the matrix provides the bulk material in the composite structure and provides means for supporting and transferring applied loads. The imbedded **reinforcing material** can be in the form of continuous or discontinuous fibers, as well as particles and flakes. The properties of the composite materials are affected by the volume fraction and size of the reinforcing material. In fiber reinforced composites, the length and orientation of the fibers play an important role in tailoring the properties of the composite material[1,2,3]. Examples of materials used as reinforcing agents are glass and carbon fibers, silicon carbide, Kevlar 49, and steel filaments.

Composite materials are most commonly classified based on the matrix material; therefore, three types of composite materials are available:
1. Metal matrix composites (MMCs)
2. Polymer matrix composites (PMCs)
3. Ceramic Matrix composites (CMCs)

1.3 Mechanical Properties of Materials

Mechanical testing is used to determine a material property or a set of properties. The most typical types of tests performed on engineering materials are uniaxial tension test, impact toughness test, flexure test, hardness tests, and creep tests. This section explains how the material properties are obtained using data generated from these testing methods. Many other tests can be performed on a material to assess its behavior, but the scope of this section will be limited to the aforementioned tests.

1.3.1 Uniaxial Tension Test

The uniaxial tension test is one of the most commonly used tests to evaluate material response to static and quasi-static loading. The test is performed on universal testing machines equipped with a load cell and a displacement transducer. The tensile test specimen, having the general shape shown in Figure 1-7, is placed in the machine and gripped from both end. One of the grips is fixed, while the other displaces until the specimen fractures. Load and grip displacement data are obtained from the load cell and displacement transducer readings, respectively. However, with the use of an extensometer that is mounted on the specimen gage section, more accurate data on the displacement of the gage can be obtained. Testing procedures and parameters, as well as specimen size, can be obtained from ASTM. For molded or machined plastics ASTM D638 standard is used, and for metals ASTM E8 standard is used. Results from the uniaxial tension test are used to calculate the elastic or tensile modulus (E), yield strength (σ_y), ultimate tensile strength (*UTS*), and ductility. The material flow model can also be developed using data obtained from uniaxial tension tests.

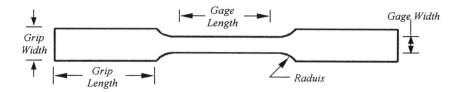

Figure 1-7. Uniaxial tensile test specimen shape

Examples of load-displacement curves for metals and plastics are shown in Figure 1-8. Figure 1-8(a) shows the load displacement curve for metals, which is characterized by a linear elastic region starting at the beginning of deformation followed by a non-linear plastic region. **Elastic deformation** is non-permanent and is recovered upon removing the applied load, but **plastic deformation** is permanent. The load reaches a peak level after which it begins to drop indicating the start of localized deformation or **necking**. Figure 1-8(b) shows load-displacement diagrams for polymers at different testing temperatures. The curves show that polymers tested at temperatures much lower than T_g are generally brittle, but their ductility increases as temperature rises. As the temperature increases to levels above T_g, a peak load is reached and a neck begins to form. However, unlike metals, the neck in polymers spreads over the gage length. As the specimen approaches it fracture point the load rises due to the stretching of molecules[2].

1.3.1.1 Tensile Modulus

The **tensile modulus** represents the material stiffness and is calculated as the slope of the load-displacement curve in the linear (elastic) region for metals, and as the initial slope of the load displacement curve for plastics. It is worth mentioning that the modulus of metals is about two orders of magnitude higher than that in polymers[3]. Also, as can be seen in Figure 1-8(b) that the tensile modulus in polymers decreases as the temperature increases.

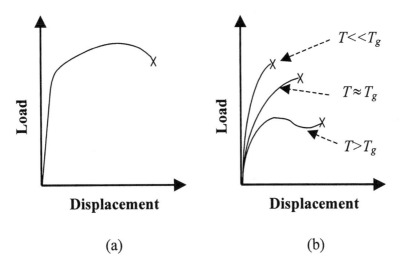

Figure 1-8. Load-displacement behavior in (a) metals, and (b) polymers[3]

1.3.1.2 Engineering Stress

The load-displacement data are converted into stress and strain data in order to obtain the material properties. **Stress** is the load bearing capacity of a material normalized per unit area. Stress is of two types: normal and shear. **Normal stress** results from an applied load that is perpendicular to the material surface and **shear stress** results from a load applied parallel to the surface. As can be seen in Figure 1-9, a load, P, applied to a surface at an angle, θ, with respect to a horizontal plane, will have a shear component, $P_s = P\cos\theta$, and a normal component, $P_n = P\sin\theta$. Normal stress is calculated by dividing P_n by the cross-sectional area ($A = w \cdot t$), as shown in Eq. 1, and the shear stress, τ, is calculated according to Eq. 2.

$$\sigma = \frac{P_n}{A} \tag{1}$$

$$\tau = \frac{P_s}{A} \tag{2}$$

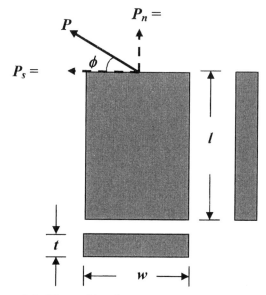

Figure 1-9. Normal load, P_n, and shear loads, P_s.

Engineering stress is the normal stress applied on a tensile specimen and is calculated by dividing the applied load by the initial cross-sectional area of the gage section of the tensile test specimen, A_o. Since the applied load in a uniaxial tension test is perpendicular to the cross-sectional area of the specimen gage section, then there is no shear component to the applied load and $P_n=P$. Tensile strength, or ultimate tensile strength (UTS), is the maximum engineering stress in the specimen and is calculated using the peak load at the onset of necking (displacement at which the load begins to drop).

Figure 1-10 shows a schematic of the original gage section size of a tensile specimen with starting length, width, and thickness labeled as l_o, w_o, and t_o, respectively. The initial cross-sectional area of the gage is calculated by multiplying the original width by the original thickness. The equation used to calculate the engineering stress is given in Eq. 3.

$$\sigma_{eng} = \frac{P}{w_o \cdot t_o} = \frac{P}{A_o} \tag{3}$$

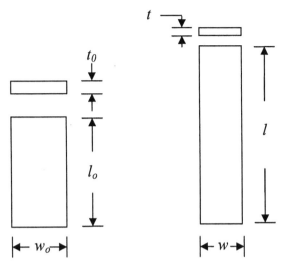

Figure 1-10. Schematic showing the original and the deformed sizes of the gage section of a uniformly stretched tensile test specimen

1.3.1.3 Engineering Strain

Upon applying a tensile load, the specimen stretches uniformly to a value of l (Figure 1-10), until the onset of necking. Accordingly, the width and thickness compress to values w and t, respectively. The initial volume of the material in the gage section, V_o, is assumed to remain constant during plastic deformation and equal to the final volume of material in the gage section, V, as shown in Eq. 4. The engineering strain, ε_{eng}, represents the change in length in the gage section of the specimen normalized per unit of initial length as shown in Eq. 5.

$$V_o = A_o l_o = Al = V$$
$$V_o = l_o \cdot w_o \cdot t_o = l \cdot w \cdot t \tag{4}$$

$$\varepsilon_{eng} = \frac{\Delta l}{l_o} = \frac{l - l_o}{l_o} \tag{5}$$

1.3.1.4 Ductility

Ductility is another material property that can be obtained from the uniaxial tension test. **Ductility** is defined as the total elongation to fracture (e_f) of the test specimen and is usually represented as a percentage based on the initial length of the gage section, l_o (Eq. 6). Uniform elongation is another measure of material ductility. **Uniform elongation** is the change in length of the tensile specimen without necking. It is calculated similar to Eq. 6, but using the length corresponding to the point of maximum load instead of the fracture length (l_f). Specimen length is typically obtained from extensometer readings.

$$e_f = \frac{l_f - l_o}{l_o} \times 100\% \tag{6}$$

1.3.1.5 True Stress and True Strain

The **true stress**, σ_{true}, represents the actual stress in the material based on the instantaneous cross-sectional area. Since the instantaneous width and thickness are not readily available, the true stress is usually calculated using engineering stress and engineering strain with the assumption that the volume of material in the gage section of the test specimen remains constant. This assumes that the sum of plastic strains representing length, width, and thickness, is equal to zero as shown from Eq. 4 and Eq. 7. Thus, the true stress is calculated according to Eq. 8.

$$\begin{aligned} \varepsilon_l + \varepsilon_w + \varepsilon_t &= 0 \\ A_o l_o &= A l \\ \frac{l}{l_o} &= \frac{A_o}{A} \\ A_o &= A(1 + \varepsilon_{eng}) \end{aligned} \tag{7}$$

$$\sigma_{true} = \frac{P}{A} = \sigma_{eng}(1 + \varepsilon_{eng}) \tag{8}$$

The **true strain**, ε_{true}, represents the change in length based on the instantaneous length. It is calculated as the natural logarithm of the ratio of instantaneous length divided by the initial length, as shown in Eq.9. The equation also shows that the true strain can be calculated using the engineering strain.

$$\varepsilon_{true} = \int_{l_o}^{l} \frac{dl}{l} = \ln(\frac{l}{l_o}) = \ln(\frac{A_o}{A}) = \ln(1 + \varepsilon_{eng}) \qquad (9)$$

1.3.2 Toughness Tests

The toughness of a material is determined using an impact test. For metals, the Charpy impact test (ASTM E23) is used to determine material toughness. For plastics, the toughness is often determined using the Izod pendulum impact resistance test described in ASTM D256. The test is similar to the Charpy impact test, except that the specimen is only supported on one end and acts as a cantilever.

A notch is used in the impact specimen in order to control the fracture process by concentrating stress in the area of minimum cross-section (the area behind the notch). It should be noted that different plastics respond differently to a given notch shape. This so called "notch sensitivity" can be evaluated using ASTM D256 test D[8].

Figure 1-11 shows a schematic of the specimen setup in a Charpy impact test. The load is applied using a pendulum that swings in the y-z plane, which is perpendicular to the notch, as shown in Figure 1-11.

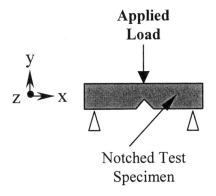

Figure 1-11. Schematic of the specimen setup in Charpy impact test

1.3.3 Hardness Tests

The material hardness is an indicator of the material resistance to deformation. Hardness can be easily measured using a flat specimen and a hardness testing machine. There are several standard hardness tests that are used to test metals including Brinell hardness (ASTM E10 for metals) and Rockwell hardness (ASTM E18 for metals and D785 for plastics). The difference between these hardness tests is related to the shape and size of the indenter, the amount of applied load, and the load application rate. The advantage of using hardness tests is that they are non-destructive and that specimen preparation is easy. The tests lead to a small indentation mark and tested parts do not need to be scrapped unless the mark affects a desirable surface appearance requirement.

For measuring the hardness of soft plastics and elastomers, a durometer may be used and the hardness is referred to as "durometer hardness" - ASTM D2240. The Shore$^{(R)}$ A or Shore$^{(R)}$ D durometer testing scales may be used for elastomers and soft plastics[6]; however, the Shore$^{(R)}$ D scale is preferred for harder materials.

The durometer is a device with a conical or a spherical indenter that is pressed into the polymer to a given level of applied pressure (depending on the test type/scale). The pressure is applied using a calibrated spring and the depth of indentation is measured. Durometer hardness readings of elastomers and plastics are sometimes followed by the indentation time. This is due to the fact that the reading may change over time due to the resilience of these materials[6].

1.3.4 Flexure Tests

The modulus of elasticity of rigid or semi-rigid plastics can also be determined using the flexure test according to ASTM D5934 standard test method. The test is a three point bending test with a load applied at a controlled rate at the mid-span between the two supports; a schematic of the test setup is shown in Figure 1-12. The applied load and the beam displacement (deflection) at the point where the load is applied are used to compute the modulus (or flexure modulus). It should be noted that the modulus calculated from flexure test data is usually slightly higher than that based on tensile test data[2].

The standard specifies the test specimen size as well as the span between the fixed support points. Testing can be performed on a universal testing machine.

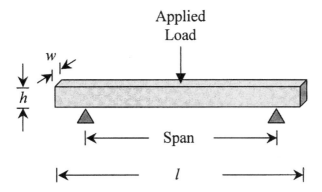

Figure 1-12. Schematic of test specimen, loading, and support in flexure tests

1.3.5 Creep Test

Creep is defined as the deformation of a material when subjected to small loads if the material is exposed to elevated temperatures for long time periods. The properties of polymers are generally sensitive to temperature and some properties are sensitive to time (as is the case with hardness testing). This indicates that polymers are likely to undergo creep deformation, and as a result their dimensional stability may be affected. Therefore, it is important to specify the maximum temperature that a material can be subjected to in service.

The **deflection temperature** of a polymer can be used as the upper limit for service temperature[2]. ASTM D648 is a testing standard for determining the deflection temperature for plastics using the flexure test. Testing is performed by applying a set load (depending on the rigidity of the tested material as specified in D648) to the test specimen, then heating the specimen at a constant rate (2 °C per minute), while the load is still applied. The test is stopped once the specimen deflection at mid-span reaches 0.25 mm. The corresponding temperature is recorded and considered to be the deflection temperature for the material[2]. This test can be used for molded material and sheet materials with thickness, $t \geq 3$ mm. If 1 mm $< t <$ 3mm, a stack of sheets with composite thickness > 3 mm can be used for testing[8].

1.4 Polymers used in Rapid Prototyping

This section focuses on the properties of thermoplastic polymers that are used in rapid prototyping. The section provides information on the properties of each type of polymer and the RP technology in which it is being used. The mechanical properties of the materials are presented in Table 1. In order for polymers to achieve their properties or to reduce cost, additives are used in polymer production. These additives include fillers (powders and fibers that can improve properties or reduce cost), plasticizers (organic chemical that improve flexibility), stabilizers (substances that help reduce degradation), and lubricants (chemicals that reduce friction and improve material flow in molds).

ABS Polymer

This polymer has an amorphous structure and is made of three monomers: Acrylonitrile (C_3H_3N), Butadiene (C_4H_6), and Styrene (C_8H_8). The combination of these monomers leads to the formation of two different co-polymer phases to make up the ABS polymer. The first phase is a hard styrene-butadiene copolymer and the second is a rubbery styrene-acrylonitrile co-polymer[1]. ABS polymers are used in several applications including automotive, consumer electronics, and appliances.
The polymer has several desirable properties such as its good strength and relatively high toughness. The properties can be manipulated by adjusting the amount of each monomer. It is used in stereolithography (STL) fused deposition modeling (FDM) and selective laser sintering (SLS) RP technologies.

Acrylics

Acrylics are polymers with an amorphous structure that are obtained from acrylic acid. They are noted for their good transparency, which allows them to transmit about 90% of incident light[4]. This makes them good candidates for replacing glass; however, they have lower scratch resistance than glass. Acrylics are available in many colors and an example is Plexiglas. They are used in automotive and optical instrument applications Acrylics are used to produce prototype parts using the STL technology.

Cellulose

Cellulose is a natural polymer with an amorphous structure and the chemical composition $C_6H_{10}O_5$. Wood, which is composed of 50% cellulose, is one of the primary natural sources of this polymer. When subjected to heat, cellulose disintegrates prior to reaching its melting temperature. Therefore, cellulose needs to be combined with other materials

in order to produce a thermoplastic with more desirable thermal behavior. Four groups of cellulose-based polymers are commercially available[4]:
1. Cellulose acetate
2. Cellulose acetate butyrate
3. Cellulose acetate propionate
4. Cellulose nitrate

Cellulose is used in laminated object manufacturing (LOM) rapid prototyping technology.

Nylon
Nylons are members of the polyamide (PA) family and have mostly crystalline structures. The most commonly used Nylons are Nylon6 (PA6) Nylon6,6 and (PA6,6). Nylons have good wear resistance and their strength can be improved by reinforcing them with glass fiber. Nylons are the main prototyping materials used by the laser sintering (LS) rapid prototyping technology and are also used by the FDM technology.

Polycarbonate
Polycarbonates are polymers with the composition of $[C_3H_6(C_6H_4)_2CO_3]_n$. They have an amorphous structure and are characterized by their good creep resistance and good toughness. The have excellent resistant to heat compared with other polymers. They are used in automotive windshield applications as well as product housings. They are used for STL prototyping applications.

Thermoplastic Polyester
Polyesters are polymers of a semi-crystalline structure and are of two types: thermoplastic polyesters and thermosetting polyesters. Two commonly used types of polyesters are polybutylene terephthalate (PBT) and polyethylene terephthalate (PET). PBT is used in the manufacture of automotive luggage racks and headlight components. PET is used by the packaging, automotive, and electrical industries due to its high toughness and temperature resistant. Polyesters are used in SLS rapid prototyping technology.

Polyethylene (PE)
Polyethylene has a semi-crystalline structure and is of the chemical composition $[C_2H_4]_n$. Polyethylene has good toughness and relatively excellent resistance to chemical attack. It can be processed by almost any thermoplastic processing method, thus it is the most commonly used type of thermoplastics.

Polyethylene has two major types: low density polyethylene (LDPE) and high density polyethylene (HDPE). Differences in properties between the two types are caused by structural differences since density is directly proportional to the degree of crystallinity (% crystalline) of the material structure. Polyethylene is used in FDM technology.

Polypropylene (PP)

Polypropylene has a semi crystalline structure with a high degree of crystallinity. PP, which has the chemical composition $[C_3H_6]_n$, is the lightest of all currently available plastics. It has good resistance to chemical attack and has properties that are comparable with HDPE. It is used in the manufacture of one-piece plastic hinges[1]. It is used in the FDM rapid prototyping technology.

Polyvinylchloride (PVC)

Polyvinylchloride has an amorphous structure with the chemical composition $[C_2H_3Cl]_n$. The rigidity of PVC is inversely proportional to the amount of plasticizer it contains. In addition, PVC contains stabilizers in order to control its instability when subjected to light and heat. PVC is used in SLS rapid prototyping technology.

1.5   Material Selection

The selection of materials for RP purposes depends greatly on the application, especially if the functional requirements of the product are to be evaluated using the prototyped part[5]. As mentioned earlier, one of the challenges in assessment of mechanical properties of a product using RP parts is that the material composition used for prototyping may not be the same as the production material. In addition, the material structure developed by RP processes, the surface finish, and dimensional tolerances will not match those generated by actual production processes[7]. Therefore, it is important to use a RP material that represents the properties being evaluated and not all the properties of the prototyping material. For example, if a product requires a certain level of stiffness at an elevated service temperature level, then the material that is selected for prototyping should have similar stiffness at this elevated service temperature. Other properties of the prototyping material may not be as significant.

According to Mueller[5], the challenge with performing functional tests on prototyping materials is that the production part may not pass functional tests that the rapid prototyped part might have passed, and vice versa. To remedy this issue, Mueller suggests using a four-step approach for selecting the proper type of rapid prototyping material:

1. Examine the service loading conditions and other significant mechanical requirements of the product.
2. Identify the relevant material properties that address these requirements.
3. Quantify the sensitivity of the design to variations in the critical material properties.
4. Select a prototyping material with mechanical properties that falls within the acceptable limits and identify a prototyping process that is capable of producing the required properties in the prototyped part.

Examples of critical properties that can be considered in the material selection process include, loading conditions, service temperature, impact resistance, stiffness and ductility, etc.

References

1. M. Groover, Fundamentals of Modern Manufacturing: Materials Processes, and Systems, 2nd edition, John Wiley and Sons, Inc. (2004).
2. J. Schey, Introduction to Manufacturing Processes, 3rd edition, McGraw Hill (2000).
3. E.P. DeGarmo, J.T. Black, and R.A. Kohser, Materials and Processes kin Manufacturing, 8th edition, John Wiley and Sons, Inc., (1999).
4. S.D. El Wakil, Processes and Design for Manufacturing, 2nd edition, PWS Publishing Co. (1998)
5. T. Muller, Truly Functional Testing: Selecting Rapid Prototyping Materials So That Prototypes Predict the Performance of Injection Molded Plastic Parts, SME Technical Paper No. TP04PUB238, (2004).
6. www.matweb.com/reference/shore-hardness.asp [accessed on July 8, 2005]
7. R. Hague, S. Mansour, and N. Saleh, Material and Design Considerations for Rapid Manufacturing, International Journal of Production Research, **42**(22), 4691-4708 (2004).
8. ASTM International Standard Number D256-05.

Chapter 2

IGES Standard Protocol for Feature Recognition CAD System

Emad Abouel Nasr, PhD Candaite[1] and Ali Kamrani, PhD[2]
Industrial Engineering Department, Faculty of Engineering, University of Houston, 213 Engineering Building 2, Houston, TX 77204-4008, [1]*emadsamir60@yahoo.com,* [2]*akamarni@uh.edu*

Abstract:
   Automatic feature recognition from CAD solid systems highly impacts the level of integration. CAD files contain detailed geometric information of a part, which are not suitable for using in the downstream applications such as process planning. Different CAD or geometric modeling packages store the information related to the design in their own databases. Structures of these databases are different from each other. As a result no common or standard structure has been developed so far, that can be used by all CAD packages. For that reason this chapter proposes an intelligent feature recognition methodology (IFRM) to develop a feature recognition system which has the ability to communicate with various CAD/CAM systems. The proposed methodology is developed for 3D prismatic parts that are created by using solid modeling package by using constructive solid geometry (CSG) technique as a drawing tool. The system takes a neutral file in Initial Graphics Exchange Specification (IGES) format as input and translates the information in the file to manufacturing information. The boundary (B-rep) geometrical information of the part design is then analyzed by a feature

recognition program that is created specifically to extract the features from the geometrical information based on a geometric reasoning approach by using object oriented design software which is included in C++ language. A feature recognition algorithm is used to recognize different features of the part such as step, holes, etc. Finally, a sample application description for a workpiece is presented for demonstration purposes.

Key words:

CAD, CAM, CAPP, CIM, Feature Recognition, IGES

2.1. History & Overview

The origins of solid modeling are traced back to the key technological inventions of the 1950s which was Computer Graphics and NC machining. These spawned the developments of computer based geometric systems to aid in the description of object's geometry, which is the main activity to design and manufacture of mechanical parts. This resulted research into the development of Computer Aided Design and Computer Aided Manufacturing (CAD/CAM). Preliminary systems used electronic drafting and wireframe models to represent the shape of three dimensional objects. Subsequent systems developed in the 1960s used polygonal and surface based models which were utilized for a variety of applications in aerospace, marine and automotive industries[4].

Until the 1970s, these models were used in a broad manner. They were merely a collection of lower dimensional entities (polygons, surfaces, lines, curves and points) put together in an unstructured manner to represent a real object. The developments in CAD/CAM led to the crucial questions about the uniqueness and the validity of these models, issues which until then were unimportant from the point of view of computer graphics and its applications. In the late 1970s, these issues were resolved by the Production Automation Project at the University of Rochester, where the term "solid modeling" was coined[11]. This group developed new mathematical models for representing solids and identified the relevant properties of an informationally complete representation. They also identified the mathematical operations that could be used to manipulate these models. After that several other models have been proposed along with different representation schemes[14].

In the 1980s, several solid modeling systems were developed and used in the commercial CAD/CAM world, including the automobile, aerospace and manufacturing industries. Moreover, many advanced CAD/CAM applications of solid modeling have emerged such as feature and constraint based modeling, automatic mesh generation for finite element analysis, assembly planning including interference checking, higher dimensional modeling for robotics and collision avoidance, tolerance modeling, automation of process planning tasks, etc. Currently, solid modeling techniques have gained importance in the industry and are also actively pursued as a research field in academic institutions[12]. Figure 2-1 lists a summary of solid modeling history evolution technology.

To summarize, solid modeling provides a framework to model and represents an object's shape in the computer, and to perform operations. In addition to, a group of application independent geometric tools and algorithms is provided which can be used to query/analyze the model to obtain unambiguous results. These tools can be used or combined with other application specific tools to perform the required task. The issues related to data structures and geometric algorithms, their efficiency, reliability and robustness also form an important aspect of solid modeling[15].

The academic effort in solid modeling utilizes several disciplines in many applications. That is including algebraic geometry and topology, differential geometry and topology, combinatorial topology, computer science, and numerical analysis. Another well established and closely related field is *Computer Aided Geometric Design* (CAGD) which concentrates on developing techniques for freeform surface design used to model curves and surfaces[10].

2.2. Standard Data Format

This section presents discussions related to a standard product data format of an object which considered as the most important tool towards the standardization of product data and at the same time towards the compatible exchange of information among various CAD and CAM systems. IGES format is addressed in details as one of the popular standard format.

2.2.1 Data Transfer in CAD/CAM Systems

Field of data transfer between different CAD/CAM systems is a well established one for a number of years and the paramount importance of CAD/CAM/CAE data transfer between manufacturers and their suppliers and subcontractors has become more apparent. In the early years of CAD/CAM industry, software packages were developed which were employed as direct translators between different systems. Obviously, these packages were used with great success. However, as the number of

CAD/CAM system vendors was increased, the impracticality of using direct translators becomes more apparent. Hence, a few number of neutral format translators developed by various organizations in different countries were introduced into the industrial market[9].

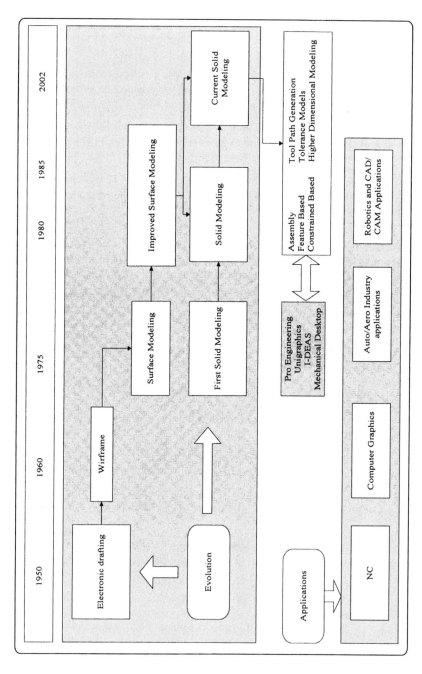

Figure 2-1. Solid Modeling technology evolutions

Some of these translators were tailor made for specific industries and others were accepted as standard tools by various authorized standard organizations. Some of these standards such as Standard for Exchange of Product data (STEP), Data eXchange File (DXF), Product Data Exchange Specifications (PDES) and Initial Graphics Exchange Specifications (IGES) have proved more popular with CAD/CAM system vendors and users[1, 8, 13]. IGES was first developed by National Aeronautical and Space Administration and National Bureau of Standards in 1979. Soon after it was adapted and recognized by American National Standard Institute (ANSI) as a standard tool format. Consequently, IGES has become an acceptable and widely used neutral format translator by many CAD/CAM system vendors[2, 3]. Even though some translators are more broadly used than IGES, this neutral format translator has been through many revisions and has proved reasonably a comprehensive tool in transferring data for parts designed by wireframe, surface or solid models. For this book, IGES version 5.3 documentation was closely studied and adopted[16].

2.2.2 Initial Graphics Exchange Specifications (IGES)

Initial Graphics Exchange Specification (IGES) was developed as a neutral data format for the transmission of CAD data between dissimilar CAD/CAM systems. Although the IGES format does not provide a suitable data format for downstream manufacturing applications, it can be considered as the major driving force to achieve the international standard of product data and the data exchange format. Therefore it is described in details in this section. In order to transfer information, translation is done from one native format to the neutral file and then to another native format[17]. As shown in Figure 2-2, the number of processors needed to transfer data among N different CAD systems using a neutral file is 2 * N.

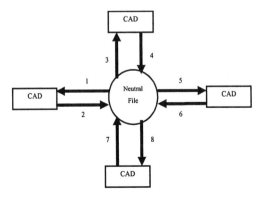

Figure 2-2. Translation using a Neutral File

IGES file is just a document that specifies what should go into a data file. Programmers should write software to translate from their system to the IGES format or vice versa. The program that translates from a native CAD format to IGES is called preprocessor. The program that translates from IGES to another target format is called postprocessor as shown in Figure 2-3.

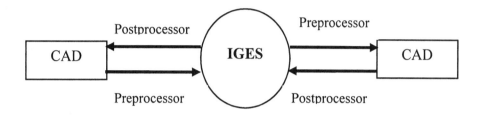

Figure 2-3. IGES translators

2.2.2.1 Structure of IGES file

Similar to the most CAD systems, IGES is based on the concept of entities. Entities could range from simple geometric objects, such as points, lines, plane, and arcs, to more sophisticated entities, such as subfigures and dimensions. Entities in IGES are divided in three categories:
1. Geometric entities: such as arcs, lines, and points that define the object.
2. Annotation entities: such as dimensions and notes that aid in the documentation and visualization of the object.
3. Structure entities: Those define the associations between other entities in IGES file.

An IGES file is a sequential file consisting of a sequence of records. The file formats treat the product definition to be exchanged as a file of entities, each entity being represented in a standard format, to and from which the native representation of a specific CAD/CAM system can be mapped. IGES file is written in terms of ASCII characters as a sequence of 80 character records.

An IGES file consists of five sections which must appear in the following order: Start section, Global section, Directory Entry (DE) section, Parameter Data (PD) section, and Terminate section, as shown in Figure 2-4. The role of these sections is summarized in the following subsections.

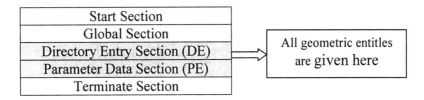

Figure 2-4. IGES file structure

2.2.2.1.1 Start Section

The Start section is a human readable introduction to the file. It is commonly described as a "prologue" to the IGES file. This section contains information such as the names of the sending (source) and receiving (target) CAD/CAM systems, and a brief description of the product being converted.

2.2.2.1.2 Global Section

The Global section includes information that describe the preprocessor and information needed by the postprocessor to interpret the file. Some of the parameters that are specified in this section are:

1. Characters used as delimiters between individual entries and between records (usually commas and semicolons respectively),
2. The name of the IGES file itself,
3. Vendor and software version of sending (source) system,
4. Number of significant digits in the representation of integers and single and double precision floating point numbers on the sending systems,
5. Date and time of file generation,
6. Model space scale,
7. Model units,
8. Minimum resolution and maximum coordinate values,
9. Name of the author of IGES file.

2.2.2.1.3 Directory Entry Section (DE)

The DE section is a list of all the entities defined in the IGES file together with certain attributes associated with them. The entry for each entity occupies two 80-character records which are divided into a total of twenty 8-character fields as shown in Figure 2-5. The first and the eleventh (beginning of the second record of any given entity) fields contain the entity type number such as 100 for circle, 110 for lines, etc. The second field contains a pointer to the parameter data entry for the entity in the PD section. The pointer of an entity is simply its sequence number in the DE section. Some of the entity attributes specified in this section are line font, layer number, transformation matrix, line weight, and color.

Column	1-8	9-16	.	49-56	65-72	73-80
Line 1	Entity Type	Parameter Entry Pointer		Transformat -ion Matrix	Visible Entity Switch	Sequence Number
Line 2	Entity Type					Sequence Number

Figure 2-5. Structure of Directory Section

2.2.2.1.4 Parameter Data Section (PD)

The PD section contains the actual data defining each entity listed in the DE section as shown in Figure 2-6. For example, a straight line entity is defined by the six coordinates of its two endpoints. While each entity has always two records in the DE section, the number of records required for each entity in the PD section varies from one entity to another (the minimum is one record) and depends on the amount of data. Parameter data are placed in free format in columns 1 through 64. The parameter delimiter (usually a comma) is used to separate parameters and the record delimiter (usually a semicolon) is used to terminate the list of parameters. Both delimiters are specified in the Global section of the IGES file. Column 65 is left blank. Columns 66 through 72 on all PD records contain the entity pointer specified in the first record of the entity in the DE section.

2.2.2.1.5 Terminate Section

The Terminate section contains a single record which specifies the number of records in each of the four preceding sections for checking purposes.

Field	1	2	3	4	5	6	7	8	..	73-80
Circle	100	Z	X (center of circle)	Y	X_1 (start point)	Y_1	X_2 (end point)	Y_2		Sequence Number
Line	110	X_1	Y_1 (start point)	Z_1	X_2 Z_2 (end point)	Y_2				Sequence Number

Figure 2-6. Structure of Parameter Data Section

2.3. The Feature Extraction Methodology

In this section, a methodology for feature analysis and extraction of prismatic parts for CAM applications is developed and presented. This approach aims to achieve the integration between CAD and CAM. Different CAD or geometric modeling packages store the information related to the design in their own databases. Structures of these databases are different from each other. As a result no common or standard structure has so far developed yet that can be used by all CAD packages. For that reason this research will try to develop an Intelligent Feature Recognition Methodology (IFRM) which has the ability to communicate with the different CAD/CAM systems.

The part design is introduced through CAD software and it is represented as a solid model by using CSG technique as a design tool. The solid model of the part design consists of small and different solid primitives combined together to form the required part design. The CAD software generates and provides the geometrical information of the part design in the form of an ASCII file (IGES) that is used as standard format which provides the proposed methodology the ability to communicate with the different CAD/CAM systems. The boundary (B-rep) geometrical information of the part design is analyzed by a feature recognition program that is created specifically to extract the features from the geometrical information based on the geometric reasoning and object oriented approaches. The feature recognition program is able to recognize these features: slots (through, blind, and round corners), pockets (through, blind, and round corners), inclined surfaces, holes (blind and through) and steps (through, blind, and round corners), etc. These features are called manufacturing information that are mapped to process planning as an application for CAM. Figure 2-7 shows the structure of the proposed methodology.

The intelligent feature recognition methodology (IFRM) presented in this chapter consists of three main phases: (1) a data file converter, (2) an object form feature classifier and (3) a manufacturing features classifier (production rules). The first phase converts a CAD data in IGES/B-rep format into a proposed object oriented data structure. The second phase classifies different part geometric features obtained from the data file converter into different feature groups. The third phase maps the extracted features to process planning's point of view. Figure 2-8 shows a basic flowchart of the proposed system. The sections that follow will describe the steps of feature extraction in details.

2.3.1 Conversion of CAD data files into Object Oriented Data Structure

As mentioned earlier, IGES is a standard file format for the data defining the object drawing in 3D CAD systems in B-rep structure. The entry fields in IGES format consist of an object's geometric and topological information. The geometric information includes the definition of lines, planes, circles, and other geometric entities for a given object, and the topological information defining the relationships between the object's geometric components, for example, in terms of loops (external loop and internal loop).

An external loop gives the location of major geometric shapes and an internal loop represents a protrusion or a depression (pocket or hole) on an external loop. The fundamental IGES entities, which are related to representing a solid in B-rep structure, are discussed in the following subsection to understand how these entities are defined[16].

2.3.1.1 Basic IGES Entities

Line (entity 110)
A line in IGES file is defined by its end points. The coordinates of start point and terminate point are included in parameter data section of this entity.

Circular Arc (entity 100)
To represent a circular arc in modeling space, IGES provides the information including a new plane (X_T, Y_T) in which the circular lies, the coordinates of center point, start point, and terminate point. A new coordinate system (X_T, Y_T, Z_T) is defined by transferring the original coordinate system (X_O, Y_O, Z_O) via a transformation matrix and all coordinates of points (center point, start point, and terminate point) related to this new coordinate system. The order of end points is counterclockwise about Z_T axis.

Transformation Matrix (entity 124)
This entity can give the relative location information between two

coordinate systems, X_O, Y_O, Z_O coordinate system and X_T, Y_T, Z_T coordinate system.

$$\begin{pmatrix} R_{11} & R_{12} & R_{13} \\ R_{21} & R_{22} & R_{23} \\ R_{31} & R_{32} & R_{33} \end{pmatrix} \times \begin{pmatrix} X_O \\ Y_O \\ Z_O \end{pmatrix} + \begin{pmatrix} T_1 \\ T_2 \\ T_3 \end{pmatrix} = \begin{pmatrix} X_T \\ Y_T \\ Z_T \end{pmatrix} \quad (2.1)$$

Where

$$\begin{vmatrix} R_{11} & R_{12} & R_{13} \\ R_{21} & R_{22} & R_{23} \\ R_{31} & R_{32} & R_{33} \end{vmatrix} = 1$$

Surface of Revolution (entity 120)

A surface is created by rotating the generatrix about the axis of rotation from the start position to the terminal position. The axis of rotation is a line entity. The generatrix may be a conic arc, line, circular arc, or composite curve. The angles of rotation are counterclockwise about the positive direction of rotation axis.

Point (entity 116)

A point is defined by its coordinates (X, Y, Z).

Direction (entity 123)

A direction entity is a non-zero vector in 3D that is defined by its three components with respect to the coordinate axes. The normal vector of surface can be determined by this entity.

Plane surface (entity 190)

The plane surface is defined by a point on the plane and the normal direction to the surface.

Vertex List (entity 502)

This entity is used to determine the vertex list which contains all the vertexes of the object.

Edge List (entity 504)

This entity is used to determine the edge list which contains all the edges of the object.

Loop (entity 508)

This entity is used to determine the loops which involved in all faces of the object.

Face (entity 510)

This entity is used to determine faces which consist of the object.

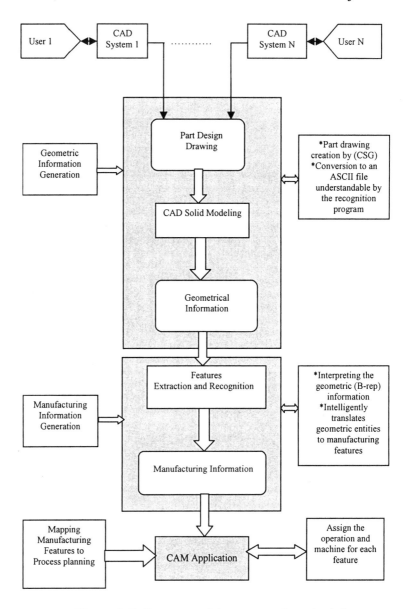

Figure 2-7. Structure of the proposed methodology

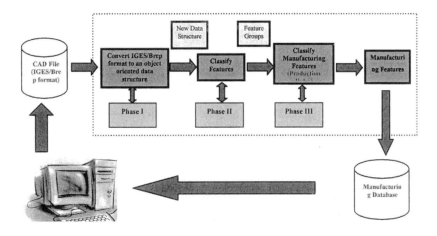

Figure 2-8. Flowchart of extraction and classification of features

Shell (entity 514)

The shell is represented as a set of edge-connected, oriented used of faces. The normal of the shell is in the same direction as the normal of the face.

Right Circular Cylindrical Surface (entity 192)

The right circular cylindrical surface is defined by a point on the axis of the cylinder, the direction of the axis of the cylinder and a radius.

2.3.2 The Overall Object-Oriented Data Structure of the Proposed Methodology

In order to have a good generic representation of the designed object for CAM applications especially for process planning, the overall designed object description and its features need to be represented in a suitable structured database. An object oriented representation will be used in this research. The first step toward automatic feature extraction will be achieved by extracting the geometric and topological information from the (IGES/Brep) CAD file and redefining it as a new object oriented data structure as demonstrated in Figure 2-9.

In this hierarchy, the highest level data class is the designed object (shell). An object consists of manufacturing features that can be classifies into form features which decomposed of either simple or compound/intersecting features. A simple feature is the result of two intersecting general geometric surfaces while compound/intersecting feature is one that results from the interaction of two or more simple features (slot and pocket) as shown in Figure 2-10. Features are further classified into concave or convex as attributes in the generic feature class. Concave features consist of two or more concave faces, and convex features are decomposed of either one or more convex faces or the interaction between other features in the object as shown in Figure 2-11.

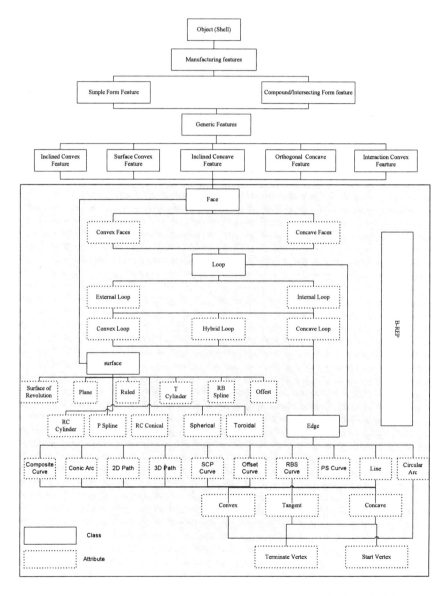

Figure 2-9. Hierarchy of classes and attributes of the designed object

Due to the attributes of the geometric entities of form features (FF), they will be classified into interior form feature ($FF_{interior}$), which is located inside the basic surface, and exterior form feature ($FF_{exterior}$), which is formed by the entire basic surface with its adjacent surfaces. The basic surface refers to the surface in which there are features located in that surface. For the interior form features ($FF_{interior}$), they can be further classified into two low-level categories, convex interior form feature ($FF_{interior\_convex}$) and Concave interior form feature ($FF_{interior\_concave}$). $FF_{interior\_convex}$ is the convex portion in a basic surface, while $FF_{interior\_concave}$ is the concave geometric portion in the surface[3].

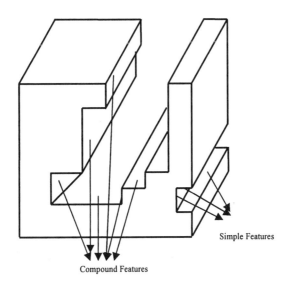

Figure 2-10. Simple and Compound Features

Examples of these form feature categories are shown in Figure 2-12. Figure 2-12 (a) shows a convex portion of the top surface investigated (basic surface) and hence this constitutes a $FF_{interior\_convex}$ (boss), while Figure 2-12 (b) shows a $FF_{interior\_concave}$ in the basic surface. A $FF_{exterior}$ is shown in Figure 2-12 (c) in which $FF_{exterior}$ is constituted by the entire basic surface and its two adjacent surfaces. The $FF_{interior\_concave}$ in Figure 2-12 (b) is a through cylindrical hole. A blind cylindrical hole or a pocket in a basic surface is also a $FF_{interior\_concave}$.

To define concave features, it is basically equivalent to identifying concave faces which are simply defined by a concave edge that joins two adjacent faces. A concave edge is determined by the concavity test that will be explained later. In general, the edge is defined by a pair of vertices described in the part drawing properties in terms of coordinates (X, Y, Z). On the other hand, convex features can be defined and classified as either inclined, interaction, or surface as shown in Figure 2-13. Inclined convex features are defined by a set of convex faces which are not parallel or perpendicular to minimal enclosing box. The second type of convex feature (interaction) results from the interaction of two or more features. Surface convex features are features that lie on the minimal enclosing box as seen in Figure 2-13.

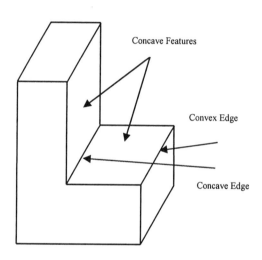

Figure 2-11. Convex and Concave Features and Edges

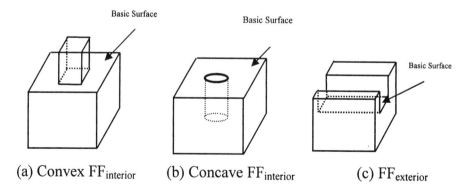

(a) Convex $FF_{interior}$ (b) Concave $FF_{interior}$ (c) $FF_{exterior}$

Figure 2-12. Classification of Interior and Exterior Form Features

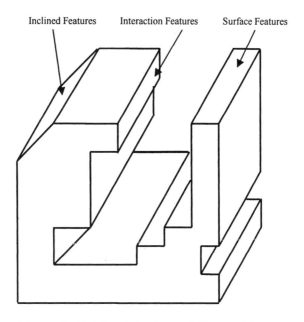

Figure 2-13. Classifications of Convex Features

2.3.2.1 Geometry and Topology of B-rep

The basic geometric entities of a 3D CAD model based on B-Rep description are vertex, edge and face. The compound entities, which consist of basic geometric entities, are shell and loops. Shells and loops are the topological entities since only topological but not geometric information is assigned to them. A solid machining object (O) model can be expressed as

$$O = (V \rightarrow v \in \text{Vertex}, E \rightarrow e \in \text{Edge}, F \rightarrow f \in \text{Faces}) \qquad (2.2)$$

where V, E and F are the sets of object vertices, edges and faces, respectively, and v, e, f are their respective elements. In this expression, each edge has two vertices and is shared by two adjacent surfaces. Each face has certain edges enclosing a specific 2D or 3D shape. The enclosed chain of edges in a surface can form one or more loops. The external loop bounds the face and the others (internal) are located inside the face.

2.3.2.1.1 Classification of Edges

Edges constitute the wireframe of a 3D solid model and they are the intersection boundaries of two adjacent faces. The following proposed edge categories facilitate the representation of the proposed methodology as proposed in Figure 2-9. To represent edge categories, the normal vectors of the two faces connected that edge and the edge direction are determined as shown in Figure 2-14. Then by applying the connectivity test that will be discussed later, the edges are classified into convex, concave, and tangent edges as shown in Figure 2-15. Also, the angle between the two faces (ɸ) is determined.

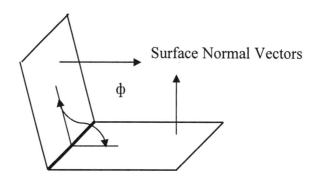

Figure 2-14. The surface normal vectors

2.3.2.1.2 Classification of Loops

A loop is the border of a surface. It is also the intersection boundary of the surface with its adjacent surfaces. In this research, a loop is used as a fundamental reference to identify the interior and exterior form features. The loop can be classified as proposed in this research into the external loop and internal loop.

External Loop is the outside boundary of a basic surface of which the loop is investigated, while internal loops are located inside the basic surface. In a basic surface, an external loop is the maximum boundary of the basic surface and the internal loop is the internal interaction boundary of the basic surface with its internal features. In Figure 2-16, there are three loops in the basic surface, one is an external loop and the other two are internal loops, which are the intersection boundaries of the basic surface with its internal convex and concave features. In addition to the loops can be further classified into other different categories as shown in Table 2-1.

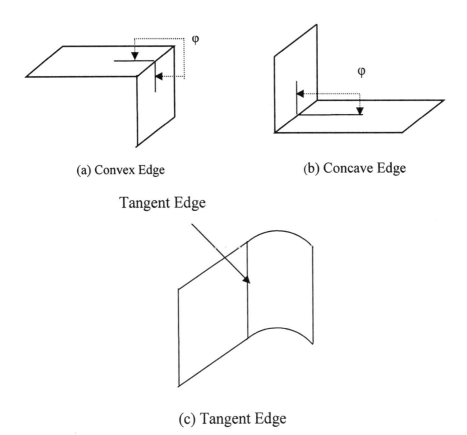

(a) Convex Edge (b) Concave Edge

(c) Tangent Edge

Figure 2-15. Classification of Edges

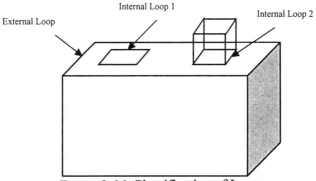

Figure 2-16. Classification of Loops

2.3.2.2 Definition of the Data Fields of the Proposed Data Structure

Generally, Faces are the basic entities that constitute the features, which are further defined by edges that are represented in terms of vertices, which are defined in terms of coordinates in CAD file. Therefore, the hierarchy of the designed object that was described in previous section (Figure 2-9) represents multilevel of different classes. All classes, except for the super class representing an object as a whole, are objects of classes that are higher up in the data structure. For example, each edge object is represented in terms of vertex objects. Table 2-2 displays the data attributes required for each class in the object oriented data structure that is defined before.

Table 2-1. Classification of Loops

Type	Definition	Case
Convex Loop (Loop class type = Convex)	All the edges are convex	If a convex loop is an internal loop, it is called internal convex loop (Internal) (Loop 2 in Fig.2-16). The external convex loop is represented as (External)
Concave Loop (Loop class type = Concave)	All the edges are concave	If a concave loop is an internal loop, it is called internal concave loop (Internal) (Loop 1 in Fig.2-16). The external concave loop is represented as (External)
Hybrid Loop (Loop class type = Hybrid)	The edges have convex and concave types	If a hybrid loop is internal loop, it is called hybrid internal loop (Internal), otherwise, It is hybrid external loop is represented as (External)

2.3.2.3 Algorithms for Extracting Geometric Entities from CAD File

The IGES file is sequentially read (on a line basis) and parsed into appropriate entry classes known as **DEntry** and **PEntry**. As the most important and useful sections of the IGES are the Directory section and the Parameter section. **DEntry** represents an entry in the directory section while **PEntry** represents an entry in the parameter section. The collection of Directory entry classes are contained in a container class called **DSection.**

Table 2-2. Definitions of classes and attributes

Class Name	Attribute	Type
Point	X_Coordinates	(Real)
	Y_Coordinates	(Real)
	Z_Coordinates	(Real)
Vertex	Inherits Point	Point
Verttex_List	Vertex_ID	(Integer)
	Vertex_Count	(Integer)
	Vertex_List	(Vector of Vertex pointers)
Edge	Edge_ID	(Integer)
	Edge_Type	(Enumerated Constants)
	Start_Vertex	(Vertex Pointer)
	Terminate_Vertex	(Vertex Pointer)
	Concavity	(Enumerated Constants)
	Face_Pointers [2]	(Array of Face Pointers)
	Loop_Pointers [2]	(Array of Loop Pointers)
	Dimension	(real)
Edge_List	Edge_Count	(Integer)
	Edge_List	(Vector of Edge Pointer)
Loop	Loop_ID	(Integer)
	Loop_Concavity	(Enumerated Constants)
	Loop_Type (External or Internal)	(Enumerated Constants)
	Edge_List	(A list of edges)
	Face_Pointers	(Pointer to face class)
Surface	Surface_Type	(Enumarated Constants)
Face	Face_ID	Number
	Surface_Pointer	(Pointer to the Surface)
	External_Loop	(Loop Pointer)
	Internal_Loop_Count	Number
	Internal_Loop_List	(Vector of Loop Pointers)
Shell	Vertex_List	(Object of Vertex_List class)
	Edge_List	(Object of Edge_List class)
	Loop_List	(Vector of Loop Pointers)
	Surface_List	(Vector of Surface Pointers)
	Face_List	(Vector of Face Pointers)
	Name	(String)
	IGES_File	(Object of IGES_File Class)
Feature	Feature_ID	(Number)
	Feature_Type	(Enumerated constants)
	Feature_Origin	(Vertex Pointer)
	Length	(Real)
	Width	(Real)
	Height	(Real)
	Redius	(Real)
	Face_List	(Vector of Face Pointers)
	Direction	(An object of Point Class)
Compound Feature	Feature_ID	(Number)
	Feature_Type	(Enumerated constants)
	Feature_List	(Vector of Feature Pointers)
	Merging_Feature	(Integer)

Similarly, the Parameter entry classes are contained in the **PSection** class. A **Parser class** object is created using these classes to parse the information present in the entries and classify the information into different classes that are used to represent different entities of the diagram described by the IGES file. Two algorithms for extraction of data from the IGES file into a proper set of data structures are defined as follows:

2.3.2.3.1 Algorithm for extracting entries from directory and parameter sections

// Algorithm to extract the directory entries
// and the parameter section entries from the iges file.
// This process takes place during the construction of an object of IGESFile class.
// Each such object represents one IGES file.
1. Create a file descriptor IgesFile
2. Create an empty dSection1 class (container to store dEntry objects).
3. Create an empty pSection1 class (container to store pEntry objects).
4. Open the Iges file for reading using IgesFile file descriptor
   // Read the file to scan and extract the directory and parameter sections.
5. While ReadLine line1 from the Igesfile
       5.1 If line1 belongs to Directory section
           5.1.1   If line1 is the first line of Dsection
               5.1.1.1 Set dIndex to 1
           5.1.2   ReadLine line2 from the Igesfile
           5.1.3   Create an object dEntry1 of class DEntry
           5.1.4   Set dEntry1 index using dIndex
           5.1.5   Initialize dEntry1 using string Line1+Line2.
           5.1.6   Add dEntry1 to dSection1 class
           5.1.7   Set dIndex = dIndex + 1
       5.2 If line1 belongs to Parameter Section
           5.2.1   If line1 is the first line of PSection
               5.2.1.1 Set pIndex to 1
           5.2.2   Create an empty string Line2
           5.2.3   while pEntry data incomplete
               5.2.3.1 ReadLine Line3 from the Igesfile
               5.2.3.2 Append Line3 to Line2
           5.2.4   Create an object pEntry1 of class PEntry
           5.2.5   Set pEntry1 index equal to pIndex
           5.2.6   Initialize pEntry1 using string Line1+Line2
           5.2.7   Add pEntry1 to pSection1 class
           5.2.8   Set pIndex = pIndex + 1

5.3   If line1 belongs to Terminate Section
  5.4   exit while loop
6 End of while loop

2.3.3.3.2 Algorithm for extracting the basic entities of the designed part

// Part-extraction module, contained in the Shell class.
// This module extracts entities and groups them into lists
// For example: it creates a list of all Faces in the object represented by the IGES file.
Procedure to extract entities.
1   Create an object vertexList1 of vertexList class
2   Create an object edgeList1 of edgeList class
3   Create a vector of pointers loop List to Loop class
4   Create a vector of pointers face List to Face class
5   Begin parsing entities from the IGES File object
    5.1   Initialize counter i=1
    5.2   For i=1 to size of dSection1 object from IGES File object
    5.3   dEntry1 = DEntry object of index i from dSection1
    5.4   if dEntry1 is a vertex list
        5.4.1   pEntry1 = PEntry object pointed to by dEntry1, obtained from pSection1 of corresponding IGES File object
        5.4.2   Initialize counter j=1
        5.4.3   For each vertex present in the pEntry1 object do
        5.4.4   Instantiate a new vertex1 object of class Vertex
        5.4.5   Assign id as j to vertex1.
        5.4.6   Initialize the object with vertex data from pEntry1
        5.4.7   Add to vertexList1 a pointer to vertex1
        5.4.8   Increment j by 1
        5.4.9   End of For loop
    5.5   End For
    5.6   Initialize counter i=1
    5.7   For i=1 to size of dSection1 object from IGES File object
    5.8   dEntry1 = DEntry object of index i from dSection1
    5.9   If dEnry1 is an edge list
        5.9.1   pEntry1 = PEntry object pointed to by dEntry1, obtained from pSection1 of corresponding IGES File object
        5.9.2   Initialize counter j = 1
        5.9.3   For each edge present in the pEntry1 class do

		5.9.4	Instantiate a new object edge1 of class Edge

- 5.9.4 Instantiate a new object edge1 of class Edge
- 5.9.5 Assign id as j to edge1
- 5.9.6 Retrieve dEntry2 object that contains edge specific data
- 5.9.7 Retrieve pEntry2 object corresponding to dEntry2
- 5.9.8 Instantiate a new object edgeS that is specific to the edge type
- 5.9.9 Initialize the edgeS object with edge1 and data from pEntry2
- 5.9.10 Assign start and terminate vertex to the edge using pointers from the vertexList1 object.
- 5.9.11 Add to edgeList object a pointer to edgeS
- 5.9.12 Increment j by 1
- 5.9.13 End For

5.10 End For
5.11 Initialize counter i=1
5.12 For i=1 to size of dSection1 object from IGES File object
5.13 dEntry1 = DEntry object of index i from dSection1
5.14 If dEntry1 is a loop
- 5.14.1 pEntry1 = PEntry object pointed to by dEntry1, obtained from pSection1 of corresponding IGES File object
- 5.14.2 Create an instance loop1 of Loop class
- 5.14.3 Assign an id to loop1 using (size of loop List + 1)
- 5.14.4 For each edge in the pEntry1
- 5.14.5 If next edge is a Vertex
  - 5.14.5.1 Add to the loop1 a pointer to the vertex
- 5.14.6 Else Add to the loop1 a pointer to the edge
- 5.14.7 End For
- 5.14.8 Add to loop List the pointer to loop1.

5.15 End For
5.16 Initialize counter i=1
5.17 For i=1 to size of dSection1 object from IGES File object
5.18 dEntry1 = DEntry object of index i from dSection1
5.19 If dEntry1 is a Face
- 5.19.1 pEntry1 = PEntry object pointed to by dEntry1, obtained from pSection1 of corresponding IGES File object
- 5.19.2 Create an instance face1 of Face class
- 5.19.3 Assign an id to face1 using (size of faceList +1)
- 5.19.4 Obtain dEntry2 containing the DEntry of the Surface type of the Face from dSection1.
- 5.19.5 Instantiate a surface1 object specific to the type of the surface of the face.

5.19.6 Add to face1 a pointer to that surface object
5.19.7 For each Loop on the Surface
  5.19.7.1 Add to face1 a pointer to loop object from loop List that represents this loop.
  5.19.7.2 For each Edge in the loop add a pointer to face1.
5.19.8 End For
5.19.9 Add to face List the pointer to face1
5.20 End For
6 End of extract entities procedure.

2.3.3.4 Extracting Form Features from CAD Files

The edge direction and the face direction are the basic entities information that is used to extract both simple and compound form features from the object data structure. The edge directions in object models can be defined such that, when one walks along an edge, its face is always on the left-hand side. When an edge is in the external loop of a face, its direction will be in a counter clockwise direction relative to the surrounding face[5, 6]. On the other hand, when an edge is in the internal loop of a face, its direction will be clockwise as shown in Figure 2-17.

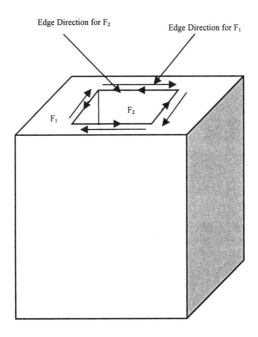

Figure 2-17. The Direction of Edge

The concave edge test used in research is based on cross product of the normal vectors of the two faces joined by a given edge. This is done by applying vector geometry to the face and edge direction vectors. Figure 2-18 shows the symbols used in this test where the ith face is designated as F_i, its corresponding normal direction vector is defined as N_i in the upward direction with respect to the given face, and the kth edge is designated as E_k.

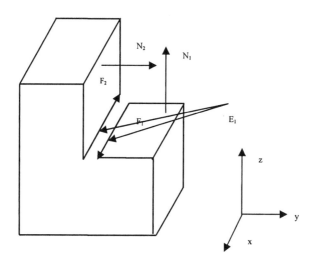

Figure 2-18. A concave Edge Example

The edge (E_K) shared by two faces (F_i and F_j) where the order is right to left from the left side of the edge view perspective. The direction vectors of the faces as described above (N_i and N_j). Finally the edge's directional vector is given with respect to face F_i using loop L_i that contains the edge (E_k). The following is the methodology for concavity test[7]:

[1] The cross (vector) product (V) of the directional vectors of the faces is determined as follows:

$$V = N_i \times N_j \qquad (2.3)$$

[2] The direction of the edge E_k with respect to the face F_i is determined. The normal vector N_i of face F_i must be the first component in the cross product of step 1.

[3] If the direction vector of edge E_k from step 2 is in the same direction of cross product V, then the edge E_k is convex edge that concludes F_i and F_j are convex faces, otherwise, it will be concave edge and F_i and F_j are concave faces. Also, if the cross product vector v is a zero vector that means the edge is tangent category.

2.3.3.4.1 An Example for identifying the concave edge/faces

The following is an example for identifying the concave edge/faces by using the example in Figure 2-18:

1. Find the face normal vectors $N_1 = [0\ 0\ 1]$ and $N_2 = [0\ 1\ 0]$.
2. Find the edge direction vector for E_1 with respect to $F_1 = [1\ 0\ 0]$
3. Find the cross product $(V) = [0\ 0\ 1] \times [0\ 1\ 0] = [-1\ 0\ 0]$
4. The edge direction of E_1 and V have the opposite direction, so the edge E_1 is a concave edge and F_1 and F_2 are concave faces.

This procedure will be done on all the edges of the object to define the concave or convex faces. Moreover, concave features will be identified by the premise that concave faces include at least one concave edge with adjacent concave faces forming a concave face group. Each concave face group defines a concave feature. Similarly, adjacent convex faces form a convex face group.

2.3.3.4.2 Algorithm for determining the concavity of the edge

The following algorithm is used to find the concavity of a line entity. This algorithm is used by the Line class to find the entities.

1   Length of line = $\sqrt{[(startX-termX)^2 + (startY-termY)^2 + (startZ-termZ)^2]}$

2   concavity = UNKNOWN
3   If face1 surface type == PLANE and face2 surface type == PLANE
      3.1.1   Assign crossDir = cross product of face1 normal vector and face2 normal vector
   3.2 If crossDir == 0
      3.2.1   concavity = TANGENT
   3.3 Else
      3.3.1   Calculate the direction vector edgeDir for the line with respect to the loop
          3.3.1.1   If crossDir is in the same direction as edgeDir
              3.3.1.1.1   concavity = CONVEX
          3.3.1.2   Else
              3.3.1.2.1   concavity = CONCAVE
4   If face1 surface type == RCCSURFACE and face2 surface type == PLANE

    4.1   Find dir1 (direction) that is orthogonal to the plane containing the edge and the axis of face1
    4.2   If dir1 and normal of face2 are orthogonal to each other
        4.2.1   concavity = TANGENT

2.3.3.4.3 Algorithms for feature extraction (production rules)

The following are some of the algorithms used for extraction of features.

Rule 1: Feature: **STEP THROUGH** (Figure 2-19)
1. For every concave edge of type Line in the edge list.
2. If the two common faces (face1 and face2) of the edge (e_1) are plane and orthogonal to each other
    2.1. if outer loop concave edge count equals 1 in both the faces
        2.1.1.   STEP THROUGH found.
        2.1.2.   Create a new StepT object and add to feature list
3. End For

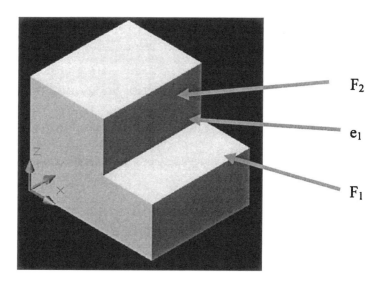

Figure 2-19. STEP THROUGH

Rule 2: Feature **STEP THROUGH ROUND CORNER** (Figure 2-20)
1. For every 3 tangent edges of type Line in the edge list (e_1, e_2, e_3)
2. If the common face of the 2 edges is quarter cylindrical surface (F_2, F_3).

2.1. The other two faces (F_1, F_4) connected to edges are perpendicular to each other.
  2.1.1. If concave edge count of the outer loops of the four faces equals 0 each.
    2.1.1.1. STEP THROUGH ROUND CORNER found
    2.1.1.2. Create a new StepT_RC object and add to feature list.
3. End For

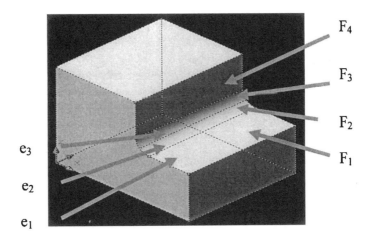

Figure 2-20. STEP THROUGH ROUND CORNER

In general, the following steps are the proposed methodology for feature's extraction and classifications:

Step 1: Extract the geometry and topology entities for the designed object model from IGES file:
  (a) Identify vertices, edges, faces, loops of the object.

Step 2: Extract topology entities in each basic surface and Identify its type:
  (a) Identify the total number of loops in each surface.
  (b) Identify the basic surface due to total number of loops.
  (c) Classify the loops into different types (concave, convex, and hybrid).

Step 3: Test the feature's existence in the basic surface based on loops.

Step 4: Identify feature type:
  (a) Identify Exterior Form Features ($FF_{exterior}$) by searching for hybrid loop.
  (b) Identify Interior Convex Form Features ($FF_{interior}$) by searching

for convex loop.
(c) Identify Interior Concave Form Features ($FF_{interior}$) by searching for concave loop.

Step 5: Identify the detailed features and extract the related feature geometry parameters:
(a) Identify feature's details (number of surfaces, surface type).
(b) Identify the parameters of each feature (length (L), width (W), height (H), radius (R).
(c) Identify the relative location of each feature due to the origin coordinates of the object.

Step 6: Identify the detailed machining information for each feature and the designed part:
(a) Identify the operation sequence of the designed part.
(b) Identify the operation type, the machine, and the cutting tool for each feature.
(c) Identify the tool approach/machining direction for each feature.
(d) Identify the removed machining volume for each feature.

2.4. An Illustrative Example

The designed object as shown in Figure 2-21 consists of five solid primitives which are four blocks (prismatic raw material, slot, blind step, and through pocket) and one cylinder (through hole). After the part is created, its IGES file will be generated. By applying the proposed methodology, the results are shown in Tables 2-3 to 2-8.

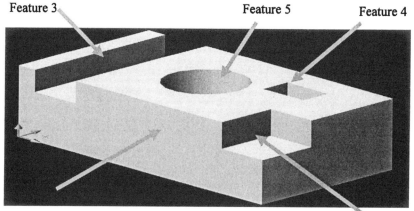

Figure 2-21. An Illustrative Example

Step 1: Extract the geometry and topology entities for the designed object model from IGES file.

Step 2: Extract topology entities in each basic surface and Identify its type.

Table 2-3. Extraction of Vertices

NO.	Vertex ID	Coordinates (X,Y,Z)	NO.	Vertex ID	Coordinates (X,Y,Z)
1	[1]	(19,8,16)	18	[18]	(25,0,16)
2	[2]	(19,8,0)	19	[19]	(30,6,16)
3	[3]	(11,8,16)	20	[20]	(30,0,0)
4	[4]	(11,8,0)	21	[21]	(30,16,0)
5	[5]	(25,13,8)	22	[22]	(30,16,16)
6	[6]	(25,10,8)	23	[23]	(7,16,11)
7	[7]	(20,13,8)	24	[24]	(7,0,11)
8	[8]	(20,10,8)	25	[25]	(2,16,11)
9	[9]	(20,10,16)	26	[26]	(2,0,11)
10	[10]	(25,10,16)	27	[27]	(2,16,16)
11	[11]	(20,13,16)	28	[28]	(2,0,16)
12	[12]	(25,13,16)	29	[29]	(7,0,16)
13	[13]	(30,6,10)	30	[30]	(7,16,16)
14	[14]	(30,0,10)	31	[31]	(0,0,16)
15	[15]	(25,6,10)	32	[32]	(0,0,0)
16	[16]	(25,0,10)	33	[33]	(0,16,16)
17	[17]	(25,6,16)	34	[34]	(0,16,0)

Table 2-4. Extraction of Edges

Edge ID	Edge Type	Starting Point	Terminate Point	Center	Length /Radius	Concavity
[1]	Line	[1]	[2]		16	Tangent
[2]	Cir. Arc	[3]	[1]	(15,8,16)	4	
[3]	Line	[4]	[3]		16	Tangent
[4]	Cir. Arc	[2]	[4]	(15,8,0)	4	

Table 2-4. Extraction of Edges (cont.)

[5]	Line	[5]	[6]		3	Concave
[6]	Line	[7]	[5]		5	Concave
[7]	Line	[8]	[7]		3	Concave
[8]	Line	[6]	[8]		5	Concave
[9]	Line	[9]	[8]		8	Concave
[10]	Line	[10]	[9]		5	Convex
[11]	Line	[10]	[6]		8	Concave
[12]	Line	[11]	[7]		8	Concave
[13]	Line	[9]	[11]		3	Convex
[14]	Line	[12]	[5]		8	Concave
[15]	Line	[11]	[12]		5	Convex
[16]	Line	[12]	[10]		3	Convex
[17]	Cir. Arc	[4]	[2]	(15,8,0)	4	
[18]	Cir. Arc	[1]	[3]	(15,8,16)	4	
[19]	Line	[13]	[14]		6	Convex
[20]	Line	[15]	[13]		5	Concave
[21]	Line	[16]	[15]		6	Concave
[22]	Line	[14]	[16]		5	Concave
[23]	Line	[17]	[15]		6	Concave
[24]	Line	[18]	[17]		6	Convex
[25]	Line	[16]	[18]		6	Convex
[26]	Line	[19]	[13]		6	Convex
[27]	Line	[17]	[19]		5	Convex
[28]	Line	[14]	[20]		10	Convex
[29]	Line	[21]	[20]		16	Convex
[30]	Line	[22]	[21]		16	Convex
[31]	Line	[19]	[22]		10	Convex
[32]	Line	[23]	[24]		16	Concave
[33]	Line	[25]	[23]		5	Convex
[35]	Line	[24]	[26]		5	Concave
[36]	Line	[27]	[25]		5	Convex
[37]	Line	[28]	[27]		16	Convex
[38]	Line	[26]	[28]		5	Convex

Table 2-4. Extraction of Edges (cont.)

[39]	Line	[29]	[24]	5	Convex
[40]	Line	[30]	[29]	16	Convex
[41]	Line	[23]	[30]	5	Convex
[42]	Line	[29]	[18]	18	Convex
[43]	Line	[31]	[28]	2	Convex
[44]	Line	[31]	[32]	16	Convex
[45]	Line	[20]	[32]	30	Convex
[46]	Line	[27]	[33]	2	Convex
[47]	Line	[33]	[31]	16	Convex
[48]	Line	[22]	[30]	23	Convex
[49]	Line	[32]	[34]	16	Convex
[50]	Line	[34]	[21]	30	Convex
[51]	Line	[33]	[34]	16	Convex

Table 2-5. Extraction of Loops

Loop ID	Loop Type	Loop Category	Face ID	Edge ID
[1]	External	Concave	[1]	[1][2][3][4]
[2]	External	Concave	[2]	[5][6][7][8]
[3]	External	Hybrid	[3]	[9][10][11][8]
[4]	External	Hybrid	[4]	[12][13][9][7]
[5]	External	Hybrid	[5]	[14][15][12][6]
[6]	External	Hybrid	[6]	[11][16][14][5]
[7]	External	Concave	[7]	[1][17][3][18]
[8]	External	Hybrid	[8]	[19][20][21][22]
[9]	External	Hybrid	[9]	[23][24][25][21]
[10]	External	Hybrid	[10]	[26][27][23][20]
[11]	External	Convex	[11]	[26][19][28][29][30][31]
[12]	External	Hybrid	[12]	[32][33][34][35]
[13]	External	Hybrid	[13]	[36][37][38][34]
[14]	External	Hybrid	[14]	[39][40][41][32]
[15]	External	Convex	[15]	[25][42][39][35][38][43][43][44][45][28][22]

Table 2-5. Extraction of Loops (Cont.)

[16]	External	Convex	[16]	[37][46][47][43]
[17]	External	Convex	[17]	[27][31][48][40][42][24]
[18]	Internal	Convex	[17]	[10][13][15][16]
[19]	Internal	Concave	[17]	[18][2]
[20]	External	Convex	[18]	[29][45][49][50]
[21]	Internal	Concave	[18]	[4][17]
[22]	External	Convex	[19]	[51][49][44][47]
[23]	External	Convex	[20]	[41][48][30][50][51][46][36][33]

Table 2-6. Extraction of Faces

Face ID	Surface Type	Normal Vector	Concavity	Number of Loops	Loop ID
[1]	Surface of revolution		Concave	1	[1]
[2]	Plane Surface (parameterized)	(0,0,1)	Concave	1	[2]
[3]	Plane Surface (parameterized)	(0,1,0)	Concave	1	[3]
[4]	Plane Surface (parameterized)	(0,0-1)	Concave	1	[4]
[5]	Plane Surface (parameterized)	(0,-1,0)	Concave	1	[5]
[6]	Plane Surface (parameterized)	(-1,0,0)	Concave	1	[6]
[7]	Surface of revolution		Concave	1	[7]
[9]	Plane Surface (parameterized)	(1,0,0)	Concave	1	[9]
[10]	Plane Surface (parameterized)	(0,-1,0)	Concave	1	[10]
[11]	Plane Surface (parameterized)	(0,0,-1)	Concave	1	[11]
[12]	Plane Surface (parameterized)	(0,0,1)	Concave	1	[12]
[13]	Plane Surface (parameterized)	(0,0,-1)	Concave	1	[13]
[14]	Plane Surface (parameterized)	(-1,0,0)	Concave	1	[14]

Table 2-6. Extraction of Faces (Cont.)

[15]	Plane Surface (parameterized)	(0,-1,0)	Convex	1	[15]
[16]	Plane Surface (parameterized)	(0,0,1)	Convex	1	[16]
[17]	Plane Surface (parameterized)	(0,0,1)	Convex	3	[17][18][19]
[18]	Plane Surface (parameterized)	(0,0-1)	Convex	2	[20][21]
[19]	Plane Surface (parameterized)	(-1,0,0)	Convex	1	[2]
[20]	Plane Surface (parameterized)	(0,1,0)	Convex	1	[23]

Step 3: Test the feature's existence in the basic surface based on loops.

Step 4: Identify feature type.

Step 5: Identify the detailed features and extract the related feature geometry parameters.

Table 2-7. Extraction of Features

Feat. ID	Feature Type	Faces ID	Edges ID	Location	Feature Name	Dimension			
						L	W	H	R
[1]	prismatic	[11][18][20] [15][18][19]	[29][30][37][40][44] [45][47][49][50][51]	[32] = (0,0,0)	Raw Material	30	16	16	
[2]	FF$_{exterior}$	[8][10][9]	[20][21][23]	[16] = (25,0,10)	Step_Blind	6	6	5	
[3]	FF$_{exterior}$	[13][12][14]	[34][[32]	[26] = (2,0,11)	Slot_Through	16	5	5	
[4]	FF$_{interior}$	[3][4][5][6][2]	[5][6][7][8] [9][12][14][11]	[8] = (20,10,8)	Pocket_Blind	8	5	3	
[5]	FF$_{interior}$	[1][7]	[1][2][3][4] [17][3][18]	(15,8,0)	Hole_Through	16			4

Step 6: Identify the detailed machining information for each feature and the designed part:

Table 2-8. Machining Information

Operation Sequence	Feat. ID	Feature Type	Operation Type	Machine	Cutting Tool	Tool Approach	Removed Volume
1	[3]	Slot_Through	Slotting_Milling	Milling	End Milling Cutter	[0,1,0] or [0,-1,0]	400.00
2	[2]	Step_blind	Shoulder_Milling	Milling	Side Milling Cutter	[-1,0,0] or [0,1,0]	180.00
3	[4]	Pocket_Blind	Pocket_Milling	Milling	End Milling Cutter	[0,0,-1]	120.00
4	[5]	Hole_Through	Drilling	Drilling	Twist Drill	[0,0,-1]	804.25

2.5. Summary

In this chapter, a methodology for feature analysis and extraction of prismatic parts for CAM applications is proposed and the implemented system is presented. This approach aims to achieve CAD/CAM integration. Different CAD or geometric modeling packages store the information related to the design in their own databases and the structures of these databases are different from each other. As a result no common or standard structure has so far been developed yet that can be used by all CAD packages. For this reason this proposed research will develop a feature recognition algorithm which has the ability to communicate with the different CAD/CAM systems.

Part design was introduced through CAD software and was represented as a solid model. The CAD software generates and provides the geometrical information of the part design in the form of an ASCII file (IGES) that is then used as standard format which provides the proposed methodology the ability to communicate with the different CAD/CAM systems. The boundary (B-rep) geometrical information of the part design is analyzed by a feature recognition program that is created specifically to extract the features from the geometrical information based on the geometric reasoning approach. The proposed feature recognition program is able to recognize the following features: slots, pockets, holes, steps, counter bore holes, counter sink holes, etc. These features, called manufacturing information, are mapped to process planning function as an application for CAM. The system is developed in C++ language on a PC-based system. Finally, a case study is used to illustrate the validity of the proposed methodology.

References

1. N. Ahmad and A. Haque, Manufacturing Feature Recognition of Parts Using DXF Files, Fourth International conference on Mechanical Engineering, Dhaka, Bangladesh, 1(1), 111-115 (2001).
2. M.P. Bhandarkar, B. Downie, M. Hardwick, and R. Nagi, Migrating from IGES to STEP: One to One Translation of IGES Drawing to STEP Drafting Data, Computers in Industry, 41(3), 261-277 (2000).
3. M.W. Fu, S.K. Ong, W.F. Lu, I.B.H. Lee and A.Y.C. Nee, An Approach to Identify Design and Manufacturing Features from A Data Exchanged Part Model, Computer Aided Design, 35(1), 979-993 (2003).
4. C. Hoffmann, and J. Rossignac, A Road Map to Solid Modeling, IEEE Transactions on visualization and Computer graphics, 2(1), 136-149 (1996).
5. J. Hwang, Rule-Based Feature Recognition: Concepts, Primitives and Implementation, Thesis, Arizona State University (1988).
6. C-H. Liu, D-B. Perng, and Z. Chen, Automatic form Feature Recognition and 3D part Recognition from 2D CAD Data, Computer and Industrial Engineering, 14(4), 689-707 (1994).
7. S. Liu, M. Gonzalez and J. Chen, Development of An Automatic Part Feature Extraction and Classification System Taking CAD Data as Input, Computers in Industry, 29(1), 137-150 (1996).
8. X. Ma, G. Zhang, S. Liu, and X. Wang, Measuring Information Integration Model for CAD/CMM, Chinese Journal of Mechanical Engineering, 16(1), 59-61 (2003).
9. S. Mansour, Automatic Generation of Part Programs foe Milling Sculptured Surfaces, Journal of Materials Processing Technology, 127(1), 31-39 (2002).
10. D. Natekar, X. Zhang, and G. Subbarayan, Constructive Solid Analysis: A Hierarchal, Geometry-Based Meshless Analysis Procedure for Integrated Design and Analysis, Computer Aided Design, 36(5), 473-486 (2004).
11. A. Reqicha and H. Voelcker, Solid Modeling: A Historical Summary and Contemporary Assessment, IEEE Computer Graphics and Applications, 2(2) 9-24 (1982).
12. O.W. Salomons, Review of Research in Feature Based Design, Journal of Manufacturing Systems, 12(2), 113-132 (1993).
13. R. Sharma and J. Gao, Implementation of STEP Application Protocol 224 in An Automated Manufacturing Planning System,

Proceedings of the Institution of Mechanical Engineers, Part B: Journal of Engineering Manufacture, **216**(1), 1277-1289 (2002).
14. S. Somashekar and W. Michael, an Overview of Automatic Feature Recognition Techniques for Computer-Aided Process Planning, Computers in Industry, **26**(1) 1-21 (1995).
15. V. Sundaarajan and P.K. Wright, Volumetric Feature Recognition For Machining Components With Freeform Surfaces, Computer Aided Design, **36**(1), 11-25 (2004).
16. B. G. William, Initial Graphics Exchange Specification IGES 5.3, ANS US PRO/IPO-100 (1996).
17. I. Zeid, CAD/CAM Theory and Practice (McGraw-Hill, Inc., 1991).

Chapter 3

The Digital Imaging and Communications in Medicine (DICOM): Description, Structure and Applications

Jinho (Gino) Lim (Ph.D.)[1] and Rashad Zein[2]
Industrial Engineering Department, University of Houston, 213 Engineering Building 2, Houston, TX 77204-4008, [1]ginolim@uh.edu, [2]rzzein@uh.edu

Abstract

This chapter is designed to introduce an exciting image data representation standard known as DICOM (Digital Imaging and Communications in Medicine) in medical imaging community. DICOM is a global information standard that is used and soon will be used by virtually every medical profession that utilizes images within healthcare industry. It is designed to ensure the interoperability of systems used to produce, store, display, send, retrieve, query, or print medical images such as computed tomography (CT) scans, Magnetic Resonance Imaging (MRI), and ultrasound. DICOM Standard has also been developed to meet the needs of manufacturers and users of medical imaging equipment for interconnection of devices on standard networks. The daily operations of DICOM Standard are currently managed by the National Electrical Manufacturers Association (NEMA). We describe a brief history, structure, current applications, and its potential use as a tool in different industries in this chapter.

Keywords:
DICOM, MRI, CT, Medical Imaging, Radiotherapy, Optimization

3.1 Introduction

3.1.1 History

The digital imaging technology has changed the way systems communicate with others using image data from simple digital pictures to complex medical images[1,2,3,7,11,13,21]. In medical community, many images have been taken to diagnose patients in hospitals since late 1970's. Such images may include computed tomography (CT) (see Figure 3-1), magnetic resonance imaging (MRI), nuclear medicine, and ultrasound.

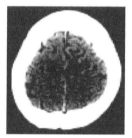 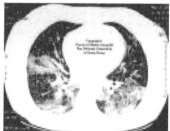

Figure 3-1. A CT image

It soon became a daily activity to transfer such image data from one location to another within the same institution or from one institution to another where different machines were used for analysis. There were two immediate needs for maintaining the data integrity during this transfer and for developing a standard for how the data should be stored. To meet this demand, the American College of Radiology (ACR) and the National Electrical Manufacturers Association (NEMA) formed a joint committee, ACR-NEMA Digital Imaging and Communications Standards Committee in early 1983. The mission of this group was to find or develop an interface between imaging equipment and whatever the user wanted to connect. Furthermore, they needed to develop standard for a dictionary of the data elements needed for proper image display and interpretation[12].

The committee launched surveys to evaluate many existing interface standards, but found none to be entirely satisfactory. However, they found useful ideas from some of the existing standards. For example, the American Association of Physicists in Medicine (AAPM) had already developed a standard format for recording images on magnetic tape[3]. The header portion would contain a description of the image along with the data elements (such as patient name) identifying it. A concept of using data elements of variable length identified with a tag or key (the name of the element) was thought to be particularly important and was adopted by the committee.

After two years of persistent work, the first version of the standard, ACR-NEMA 300-1985 (also called ACR-NEMA Version 1.0) was distributed at the 1985 RSNA (*Radiological Society of North America*) annual meeting and published by NEMA. As with many first versions, errors were found and improvements were suggested. The committee had empowered a Working Group (WG)[12,26] to follow up and improve on the standard after it was published. This Working Group answered many concerns from potential developers and began working on changes to improve the standard. In 1988, ACR-NEMA Version 2.0 was released to the market. The new version used essentially the same hardware specification and structures as Version 1.0, but it added new data elements and fixed a large number of errors and inconsistencies.

Version 2.0 provided many users with the necessary tools to interface between imaging devices and a network. However, one of the problems associated with this version was a lack of features for robust network communication. For instance, one could send a device a message that contained header information and an image, but one would not necessarily know what the device would do with the data[16]. This empowered the Working Group to make essential changes to the standard. While this issue was considered doable, it was constrained by the fact that the committee had already adopted the idea that future versions of the ACR-NEMA Standard should retain compatibility with the earlier versions. After many brainstorming sessions, it was decided that developing an interface for network support would require more than just adding elements to Version 2.0. Instead, the entire design process had to be re-engineered with the concept of object-oriented design. Furthermore, the Working Group examined types of services that were needed to communicate over different networks that allowed a top layer of communication (the application layer) to talk to a number of different network protocols[12]. These protocols are modeled as a series of layers, often referred to as "stacks" (i.e. Version 2.0 had a stack that defines a point-to-point connection.) Two other common protocols were the Transmission Control Protocol/Internet Protocol (TCP/IP) and the International Standards Organization Open Systems Interconnection (ISO-OSI)[5]. Figure 3-2 shows a diagram of the communication model developed. The basic design was that a given medical imaging application could communicate over any of the stacks to another device as long as they used the same kind of stack. With adherence to the standard, it would be possible to switch the communications between stacks without having to rewrite the computer programs of the application. This last version is what we call "DICOM standard" nowadays[12].

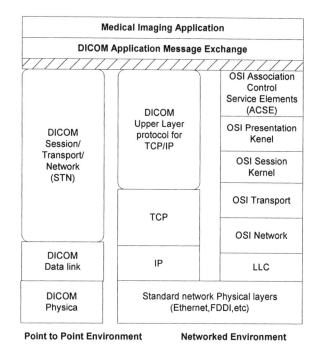

Figure 3-2. The DICOM communications protocol model.

The communication model, in general, specifies the services and the upper layer protocols that are needed to support the communication among all DICOM applications. The main task here is to ensure that the communication among DICOM applications is performed efficiently and effectively with a high level of coordination. In fact, these communication services are a subset of the services offered by the OSI Association Control Service Element (ACSE) and they are referred to as the *upper layer service*, which allows peer applications to establish associations, transfer and terminate messages. This definition of the *upper layer service* specifies the use of the DICOM *upper layer protocol* in conjunction with TCP/IP transport Protocols. The latter serves as a general purpose communication protocol and not necessarily to the DICOM standards.

Furthermore, the messages established need to be passed on to lower layers of the communications model for a communication to take place. Currently, TCP/IP and the ISO-OSI protocols are supported in DICOM, but the expansion to other protocols is relatively easy as well. Once out of the DICOM upper layer, the remainder of the communications protocol (either TCP/IP or OSI) follows the existing standards.

3.1.2 The scope of current DICOM standards

The DICOM Standard is a product of the DICOM Standards Committee and its many international working groups. It defines international standards for communication of biomedical diagnostic and therapeutic information in disciplines that use digital images and associated data. Although it has many applications outside the healthcare environments, the goals of DICOM are to achieve compatibility and to improve workflow efficiency between imaging systems and other information systems within the healthcare scope.

DICOM is a cooperative standard. The current structure is adopted by hospitals, clinics, imaging centers and specialists. Every major diagnostic medical imaging vendor in the world has incorporated the standard into their product design and most are actively participating in the enhancement of the Standard. There is a need for a continuous effort to improve the DICOM structures and potential uses.

DICOM is used or will soon be used by virtually every medical profession that utilizes images within the healthcare industry; breast imaging, cardiology, dentistry, endoscopy, mammography, oncology, ophthalmology, orthopedics, pathology, pediatrics, radiotherapy, radiology, surgery, etc. DICOM is even used in veterinary medical imaging applications. Furthermore, it will be required by all Electronic Health Record (EHR) systems that include imaging information as an integral part of the patient record. DICOM is also an integral part of Integrating the Healthcare Enterprise (IHE), which is an initiative to help both users and vendors develop approaches for integrating various medical imaging and information systems.

3.2 DICOM structure

The purpose of this section is to introduce a very basic DICOM structure; Entity-Relationship (E-R) models, DICOM components, and DICOM format. Full descriptions can be found from NEMA's website (http://DICOM.nema.org).

First, E-R models are described to show the internal interaction of DICOM entities. Second, we discuss different components of the current DICOM version. A special emphasis is given to *Conformance* because it is a critical component for a DICOM file to be compatible regardless of version of devices that a user currently has. Finally, we illustrate the format of a single DICOM file.

3.2.1 Entity- Relationship Models

The importance of modeling arises because one wants to know the context of the information when considering network communications. In a point-to-point environment, the user will know exactly what devices are connected. However, in real world applications, hundreds of devices may be attached to networks and some devices may be reconfigured dynamically to handle different data loads or tasks. This means that it may not always be possible to know how all such communications take place among the devices. Therefore, the devices may need to be re-established to build the communications that are necessary to perform the task the user has requested.

DICOM is different from ACR-NEMA Versions 1.0 and 2.0 in several major respects. ACR-NEMA relied on an implicit model of the information that is used in radiology. The data elements in those versions were grouped based on the experience of the designers. Unfortunately this led to neither sophisticated nor powerful structure. Unlike its predecessors, DICOM relies on explicit and detailed models that show the interactions among the DICOM "elements" (patients, images, reports, etc.) These are called entity-relationship models, see Figure 3-3. The advantage of these models is that they show how different required elements can interact in a given scenario. Arrows are used in the diagram to indicate the precedence relationship. The first task of the E-R model is to define the interface requirements between a picture archiving and communications system (PACS) network, and a hospital or radiology information system (HIS or RIS)[14,18]. This definition process requires that the operations in radiology be properly modeled to ensure that necessary data will be readily available for both HIS or RIS and the PACS at the same time.

Rapid Prototyping: Theory and Practice 69

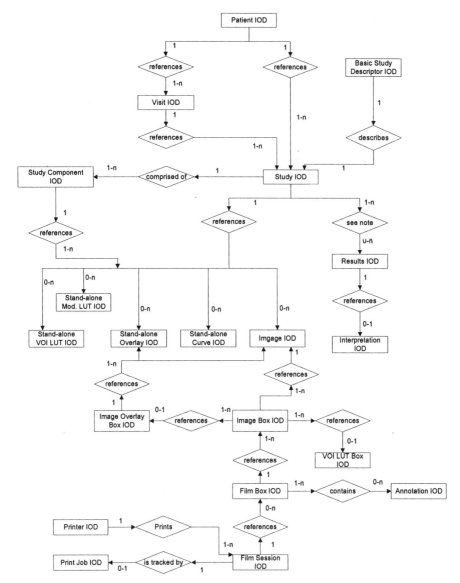

Figure 3-3. E-R Model[10]. *The rectangular boxes represent the entities which singly, or in combination, form the information objects. The diamond-shaped boxes are the relationships. The arrows represent the connections between entities and relationships, The 1's and N's indicate whether the relationship involves one entity to one entity, one to N, N to one, or N to N entities. IOD - information object definition, VOI - value of interest, LUT - look-up table, Mod – modality.*

3.2.2 DICOM Components

The current version of DICOM consists of nine parts[12]. The interrelationships of the DICOM parts are not always readily apparent. Figure 3-4 shows the current components and proposed extension of DICOM. The left hand side portion represents the parts that define network and point-to-point DICOM communications. The right hand side portion shows the parts that support communications using removable storage media. Note that some parts (parts 1, 2, 3, 5, and 6) are used in both environments while others are specific to the communications domain.

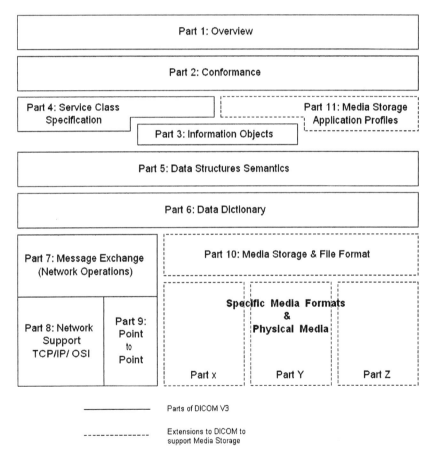

Figure 3-4. DICOM Components

Part 1 in Figure 3-4 is the document that provides an overview of the rest of the DICOM standard. It provides a description of the design principles, and definition of many terms used, and a brief description of all the other parts.

Part 2 defines a conformance to DICOM[6]. This conformance statement describes the possible choices for information objects, service classes, roles, and data encoding needed in any process to meet the required conformance. All implementations claiming DICOM conformance must be accompanied by a properly structured conformance statement. Two such statements can include (1) vendor conformance statement that specifies all device features and (2) user conformance profile (UCP) that specifies minimum system requirements. Such statements allow users to determine which components are supported by a specific implementation and compare implementations from different vendors. DICOM offers a number of building blocks or classes and requires manufacturers to describe unambiguously how their products conform to DICOM[21]. Without a conformance statement, a system does not comply with the standard. ACR-NEMA Versions 1.0 and 2.0 lacked such a mechanism so that it was possible for two devices claiming to be conformant to have different implementations and thus could cause communication problems.

Part 3 describes how information objects are defined. It then extends to define the information object classes used in DICOM. During the development of information object definitions (IODs), it was found that many users used groups of similar attributes. These were then collected together as a series of common modules that can be used by more than one IOD. Part 4 of DICOM contains the service class specifications. Service classes are built up from a set of operation primitives operating on IODs. The services can be thought of as the operations performed on the information objects. This part also contains the expected behavior of any service class. This allows implementers and users to understand what is expected from a device that supports a particular service class.

Part 5 defines the "language" that two devices will use for communication. The mechanics of the communication are defined by the message exchange protocol (part 7), and the subject matter and what needs to be done are defined by the information objects and service classes (parts 3 and 4). Part 6 of DICOM is the complete listing of all data elements along with their numeric names (or tags), their text names, and their representation (text, floating-point number, etc.).

Part 7 defines the basis of the command stream as part 5 defines the basis of the data stream. Part 8 defines the network support for exchanging the DICOM messages. For example, after a message is being constructed in part 7, it needs to be passed on to the lower layers. Part 8 facilitates this transmittal. Part 9 of the standard describes the updated version of a point to point interface. In effect, a manufacturer could choose one of the network protocols of part 8, or the point-to-point protocol of part 9, and run the same application software in either situation.

Furthermore, parts 10 and 11 address the way DICOM can use files on removable media (e.g., disk and tape) for information exchange.

3.2.3 DICOM format

DICOM is the most common standard for receiving scanned images from a hospital. Neuroimagers and neurophysiologists who wish to understand and utilize these images will need to convert the files to a readable text format[22]. Many free software are available for download fort this such as *MRIcro* (www.MRIcro.com) and *JDICOM* (www.jdicom.com).

A single DICOM file contains both a header and the image data. The header is used to store information about the patient's name, the type of scan, image dimensions, etc, while the image data contains three dimensional information of the geometry. Note that the DICOM image data can be compressed to reduce the image data size. The data compression can be done using variants of the JPEG format as well as a Run-Length Encoding (RLE) format, which is identical to the packed-bits compression found in certain TIFF format images[17]. This feature is particularly useful for electronic image data transfer among users, which is very common today.

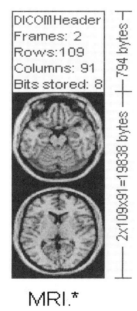

Figure 3-5. A DICOM image file[22]

Figure 3-5 shows a graphical representation of a DICOM image. In this file, the first 794 bytes are used for a DICOM format header, which describes the image dimensions and retains other text information about the scan. The size of this header depends mainly on how much header information is stored. Here, the header defines an image which has $109 \times 91 \times 2$ voxels, with a data resolution of 1 byte per voxel (so the total image size will be 19,838 bytes). The image data follows the header information (the header and the image data are stored in the same file).

We show a more detailed list of the DICOM header as displayed by MRIcro in figure 3-6. Note that DICOM requires a 128-byte preamble (these 128 bytes are usually all set to zero), followed by the letters 'D', 'I', 'C', 'M'. This is followed by the header information, which is organized in groups. For example, the group 0002hex contains 3 elements: one defines the group length, one stores the file version, and the third stores the transfer syntax.

```
First 128 Bytes: Unused by DICOM format
Followed by the characters 'D','I','C','M'
This preamble is followed by extra information e,g:
0002 0000    File Meta Elements Group Len:132
0002 0001    File Meta Info Version:256
0002 0010    Transfer Syntax UID:1.2.840.10008.1.2.1
0008 0000    Identifying Group Length:152
0008 0060    Modality: MR
0008 0070    Manufacturer: MRIcro
0018 0000    Acquisition Group Length:28
0018 0050    Slice Thickness: 2.00
0018 1020    Software Version:46\64\37
0028 0000    Image Presentation Group Length :148
0028 0002    Samples Per Pixels:1
0028 0004    Photometric Interpretation:MONOCHROME2
0028 0008    Number of Frames:2
0028 0010    Rows:109
0028 0011    Columns:91
0028 0030    Pixel Spacing:2.00\2.00
0028 0100    Bit Allocated:8
0028 0101    Bits Stored:8
0028 0102    High Bit:7
0028 0103    Pixel Representation: 0
0028 1052    Rescale Intercept:0.00
0028 1053    Rescale Slope: 0.00392157
7FED 0000    Pixel Data Group Length:19850
7FED 0010    Pixel Date:19838
```

Figure 3-6. A DICOM header as displayed by MRIcro

Multi-frame's 3D analysis tools provide an exceptionally powerful and flexible means of creating, analyzing, and examining results for any structure. DICOM multi-frame supports for cardiac angiography, digital angiography as one of its important applications. However, not all DICOM viewers support multi-frame 3D images. Figure 3-6 shows an example when such feature is supported (see the element 0028 0008 Number of Frames: 2)

In general, DICOM images can be compressed by the common JPEG compression scheme to reduce data size. Element 0002:0010 in Figure 3-6 contains extra information *Transfer Syntax UID (Unique Identification)*, which is in Figure 3-7. *Transfer Syntax UID* tells the user how the data is compressed. For example, 1.2.840.10008.1.2.4.67 in *Transfer Syntax UID* indicates that Lossless JPEG compression is used.

Transfer Syntax UID	Definition
1.2.840.10008.1.2	Raw data, Implicit VR, Little Endian
1.2.840.10008.1.2.x	Raw data, Eplicit VR **x** = 1: Little Endian **x** = 2: Big Endian
1.2.840.10008.1.2.4.**xx**	JPEG compression **xx** = 50-64: Lossy JPEG **xx** = 65-70: Lossless JPEG
1.2.840.10008.1.2.5	Lossless Run Length Encoding

Figure 3-7. A sample of a Transfer Syntax UID

In addition to the Transfer Syntax UID, the image has to be specified by some other elements such as *Samples Per Pixel* (0028:0002), *Photometric Interpretation* (0028:0004), the *Bits Allocated* (0028:0100). For most MRI and CT images, the photometric interpretation is a continuous monochrome (e.g. typically depicted with pixels in grayscale)[22]. In DICOM, these monochrome images are given a photometric interpretation of 'MONOCHROME1' (low values=bright, high values=dim) or 'MONOCHROME2' (low values=dark, high values=bright). However, many ultrasound images and medical photographs include color, and these are described by different photometric interpretations (e.g. Palette, RGB, CMYK, YBR, etc).

3.3 Current Applications

In the 1980's, it became clear to radiologists and the manufacturers of medical imaging equipment that the tremendous growth in image acquisition systems, display workstations, archiving systems, and hospital-radiology information systems have made topics like the connectivity and interoperability of all pieces of equipment an indispensable need. In order to simplify and improve equipment connectivity, medical professionals joined forces with medical equipment manufacturers in an international effort to develop DICOM. When DICOM is built into a medical imaging device, it could be directly connected to another DICOM-compatible device, eliminating the need for a custom interface[24]. DICOM defines such interface which is demonstrated in Figure 3-8.

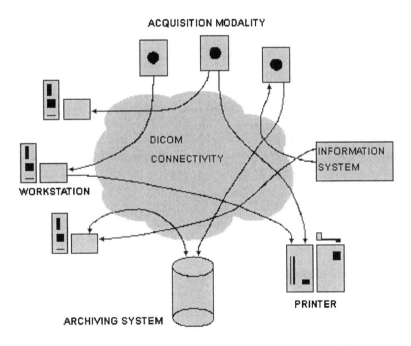

Figure 3-8. The way DICOM defines the interface[24]

DICOM was initially developed to address connectivity and interoperability problems in radiology, but today there are sections of the DICOM standard which define objects for many other modalities, including ultrasound, X-ray, and radiotherapy. The DICOM standard currently supports the following capabilities and applications[18,24]:

- **Network Image Transfer:** Provides the capability for two devices to communicate by sending objects such as images or DICOM RT (DICOM extension for radiotherapy), querying remote devices, and retrieving these objects. Network transfer is currently the most common connectivity feature supported by DICOM products.

- **On-Line Imaging Study Management:** Provides medical devices with the network capability to integrate with various information systems (HIS, RIS, archives, etc).

- **Network Print Management:** Provides the capability to print images on a networked camera. An example of this is multiple scanners or workstations printing images on a single shared camera.

- **Open Media Interchange:** Provides the capability to manually exchange objects (images or RT objects) and related information (such as reports or filming information). DICOM standardizes a common file format, a medical directory, and a physical media. Examples include the exchange of images for a publication and mailing a patient imaging study for remote consultation.

Radiology

In radiology, many problems occurred when it was not possible to support transferring large amount of image data among the radiology stations via network. The size of the image data was so large that transferring data through a network became impractical. As a consequence, it was very difficult for researchers to utilize their time effectively when they had to deal with such image data. Moreover, if there were two requests or more on a certain station for a certain image file, the system was not able to respond on a timely manner for these requests. Hence there was a need for an international standard that would understand image data and transfer this data into simple format that is easy to deal with. Introducing DICOM has made it possible for these workstations to transfer image date into text format with high level of accuracy and most importantly with a smaller size. Researchers have showed a big interest in this format as they are able now to extract correct patient demographics and image specifications. The latter information is to be stored for in a small size format for future uses.

DICOM RT

DICOM RT is a new DICOM application in radiology. It is used for radiotherapy. The following is a list of DICOM RT features that are currently recognized[25]:

- RT Structure set – Information related to anatomy such as isocenters and markers.
- RT Image – Radiotherapy images such as Digitally Reconstructed Radiographs (DRR's), Simulation images, and Portal images.
- RT Plan – Geometric and dosimetric data such as external beam components.
- RT Dose – Dose data such as reference points and isodose curves.
- RT Treatment Record – Historical record of all treatment data such that the process may be re-created at any given point in time.

Surgery

DICOM has extensively been used as a tool in Visible Light Video Sequences[20]. This concept applies to surgery and Intra Body Diagnosis. The availability of DICOM has replaced the usage of analogue tapes, which record the operation reporting. DICOM makes it easier to archive on DVD or even on servers without disturbing the flow on workstations. The usage of digital systems and servers minimize the retrieval time of operational reports extraction.

Dentistry

The applications of DICOM have crossed many barriers to reach the dental field[20]. For example, MiPACS Dental Enterprise Viewer is a comprehensive program that utilizes DICOM features for integrating seamlessly with all of the most dental popular sensors, Plate Scanners and Intra-oral Cameras (Figure 3-9). This DICOM-based Viewer offers all the tools needed to accurately diagnose and to improve the quality of images scanned, captured, or imported. One interesting feature is the ability to view similar historical series for the purpose of comparison. DICOM also supports information retrieval and display, and serves as an interface between mouth X-rays machines and MiPACS viewers. Another MiPACS application that utilizes DICOM is its Storage Server. Storage Server is an image repository and management system for most DICOM compliant modalities and client applications. DICOM gives the ability to receive captured images from any storage workstation or modality, support the Query/Retrieve process. This helps physicians retrieve patient's image data quickly and reliably.

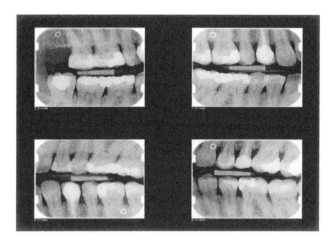

Figure 3-9. DICOM images in the dental field

Image Compression

Most of the DICOM applications outside the medical field focus on providing a mechanism for supporting the use of JPEG Image Compression through the Encapsulated Format. Moreover, DICOM provides another mechanism for supporting the use of Run Length Encoding (RLE) Compression which is byte orient.

Overall, DICOM is used by virtually every medical profession that utilizes images within the healthcare industry. These include endoscopy, mammography, opthamology, orthopedics, pathology and even veterinary medical imaging applications.

3.4 Use of DICOM in Radiation Treatment Planning

The National Cancer Institute reported that well over a million new cancer cases have been diagnosed in the United States annually recent years. More than half of these cancer patients are treated by radiation therapy. Despite the endless efforts and sophisticated treatments, many patients that are considered curable do in fact suffer from unintended side effects or even worse, die of cancer. Therefore it is vital to use very precise scientific methods to develop treatment plans that deliver high dose radiation on the tumor region while sparing nearby healthy tissues.

Cancer treatment starts from taking a series of MRI or CT scans to identify the size and the location of the tumor. The data is often stored in DICOM format to display the treatment volume for diagnosis and to develop the best possible treatment plan for the patient. Mathematical optimization has proven to be very useful for developing precise treatment plans[8,9,15,19,23]. However, optimizers often work with text data for the treatment volume geometry (tumor and healthy organs nearby) and reference radiation dose distribution to develop a treatment plan. Therefore, there is a need to extract DICOM data into a text format so that offline research can be performed to improve treatment plans. Once a treatment plan is obtained in a non-DICOM environment, one must convert the results back into DICOM format to display the results in a commercial treatment planning system for validation. A user can verify the results visually by using commercial or non-commercial software.

We now briefly describe steps how DICOM is used for designing radiation treatment plans by utilizing a given set of image data (MRI or CT scan), which was obtained from a cancer patient. There are three steps; scanning images, treatment planning, and validation of the solution and implementation.

Step 1: IMAGES in DICOM format

Using any DICOM viewer such as DICOMWORKS (Figure 3-10) that is compatible with the given image data, one can obtain many information that is associated with the treatment plan by using multiple annotation tools (arrow, text, ROI, Point Of Interest, Freehand ROI, Angles, etc.) These annotations do not alter the image but rather can be used to define the region of interest. Another useful data can be obtained in text format. These data (modality, image parameters, etc.) are essential when we apply optimization techniques for developing treatment plans.

There are many tools to view DICOM images. One of them is a computational environment for radiotherapy research (CERR) developed by the University of Washington[4]. CERR allows users to view DICOM images in Matlab. It also has a feature to reveal the text format attributes that the image contains (see Figure 3-11). This feature enables users to extract necessary information to construct such image data into a text format using any programming languages. Furthermore, Matlab has commands that can read (*dicomread*), write (*dicomwrite*) DICOM images, and display a text file containing DICOM data dictionary (*dicom.dict.txt*). Programming languages such as C/C++/Java can be very useful due to its features such as code reutilization, robustness, and extensibility. However it does not have a DICOM library as in the case of Matlab.

80 Rapid Prototyping: Theory and Practice

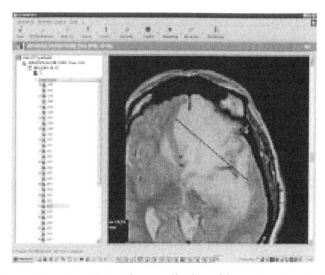

Figure 3-10. A DICOM image displayed in DICOMWORKS

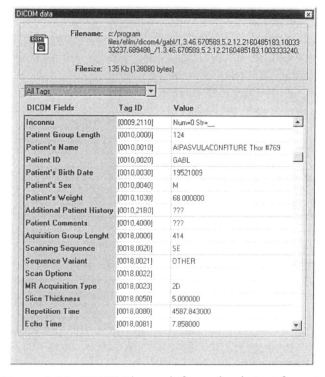

Figure 3-11. DICOM image information in text format.

Step 2: Development of a treatment plan using optimization tools.

The objective of treatment planning is to achieve desirable tumor control by planning a significant total dose of radiation to the cancerous region to sterilize the tumor without damaging the surrounding healthy tissues. An ideal treatment plan would be one that delivers radiation just right on the tumor region while surrounding tissues receive no radiation at all. We all know that it is impossible in practice due to the physics of delivering radiation from an external beam source to the tumor as well as the proximity of health tissues to the tumor. Typically, there are several clinical *targets* that we wish to irradiate, and there are several nearby organs, called *organs at risk*, that we wish to spare. Note that we usually treat targets which contain known tumors as well as regions which contain the possibility of disease spread or account for patient motion.

If we were to treat a patient with a single beam of radiation, it may well be possible to kill all the cells in the targets. However, it would also risk damaging normal cells located along the path of the beam. Therefore, beams are delivered from a number of different orientations around the patient so that the intersection of these beams includes the targets. Due to the complexity of achieving the treatment objective, this task is done by solving high level optimization problems. Several mathematical problems arise in order to develop an optimal treatment plan. Typical optimization problems are designed to answer (1) which beam angles should we include for the treatment (beam angle optimization), (2) how many beams to use and how to determine such beams shapes (aperture shape generation), (3) how long to deliver the radiation for, and (4) what is the fastest way of solving all such optimization problems (computational algorithms).

Step 3: Visualizing optimization results in DICOM interface.

CERR can display the radiation dose distribution from three different angles; axial, coronal, and saggital. It has the capability of importing and displaying treatment plans from a wide variety of commercial or academic treatment planning systems. CERR provides many clinically useful features such as dose-volume histogram, radiation dose distribution, treatment planning comparisons, etc.

Once we obtain the optimal set of beam angles and their radiation intensity levels, the results can be visualized. However, resulting dose distribution needs to be stored in DICOM format to be accessible for future use. Figure 3-12 shows dose distribution of a prostate cancer data in axial view. The screen also provides results for two other viewing angles (saggittal and coronal) in the right side panel.

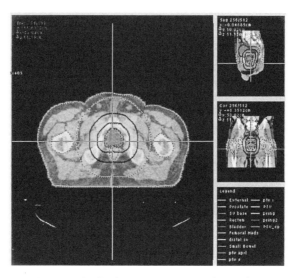

Figure 3-12. Displaying a treatment plan using CERR

Quality of a treatment plan is often measured by a dose volume histogram (DVH). DVH explains what fraction of an organ receives certain amount of radiation. In figure 3-13, more than 50% of rectum receives 40Gy or less radiation.

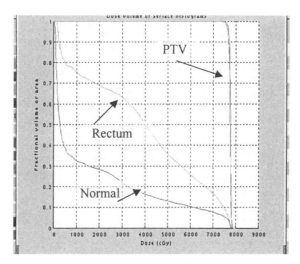

Figure 3-13. Dose volume Histogram (DVH)

Physicians will determine if such plans are acceptable for the treatment. If necessary, a better treatment can be developed by adjusting optimization parameters.

3.5 Potential use of DICOM as a tool in various industries

Potential benefits of DICOM have not well studied yet. But we know that DICOM applications are not limited to the healthcare industry. Even better yet, DICOM can play a significant role for healthcare professionals to work with others in different business sectors such as traditional manufacturing industry. For example, Rapid Prototyping (RP) is a very promising manufacturing application for developing new products in a manufacturing environment. A three dimensional (3D) object can be easily constructed using RP technique. This technique can be used to construct or develop a replacement for fractured human bones such as a part of skull. DICOM images can play a key role for developing such prototypes.

Another example of RP can be a coloring scheme for 3D organs from clinical DICOM images. This will contribute endlessly to the simulation of surgical treatment, technical training and medical education. It mainly depends on DICOM network retrieval and truly interactive 3D visualization of volumetric image data using the latest in 3D graphics technology. The application is often accompanied by a user friendly tools that allow the editing and the conversion of 3D data into Rapid Prototyping files for real and Virtual Reality applications. Some other features fall into the following categories: 1) DICOM client "query and retrieve" network functionality, 2) Image import utility for generic raw formats, 3) Simultaneously coordinated display of transverse, coronal, and sagittal views of data, 4) Truly interactive 3D visualization - pan, zoom, rotate, and 5) Automatic Rapid prototyping build optimization. Figure 3-14 shows an example of DICOM for constructing human skull using Rapid Prototyping [www.anatomics.net].

The DICOM Standards Committee has an active liaison to International Standard Organization (ISO)'s TC 215. In fact, a joint DICOM/ISO working group produced a new standard enabling Web Access to DICOM Objects. The DICOM Standard is expected to become an ISO reference standard by the end of 2005. It has been a European Standard (EN) for years[26].

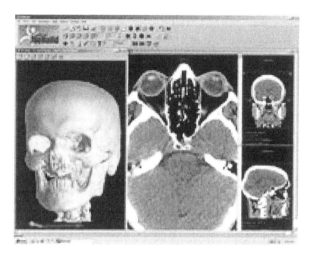

Figure 4-14. DICOM applications in rapid prototyping

3.6 Summary

As we discussed in this chapter, DICOM carries a relative long history. Despite the fact that DICOM can bring many potential benefits for imaging community, only limited groups of people have used the standards. We have identified its applications from medical imaging to rapid prototyping in a manufacturing environment. DICOM is a vendor independent standard because it facilitates interoperability by specifying (1) a set of protocols to be followed by devices, (2) syntax and semantics of protocol commands and associated information, and (3) a conformance statement which can be used as guideline by users and suppliers of equipment. Note, however, that DICOM is not an implementation specification; it does not specify either the exact format of a file or implementation details; it cannot guarantee interoperability of implementations; it does not specify testing/validation procedures to evaluate implementations.

DICOM standards committee is working hard to make continuous improvements to meet the current needs from customers. This will eventually help recognize the potential benefits of DICOM that have not been addressed anywhere. Interested readers are encouraged to visit NEMA's website (http://dicom.nema.org) for more information.

References

1. Bakker AR, Lodder H, Kouwenberg, JML., Traffic between PACS and HIS/RIS; data or information? Proceedings SPIE, 1234:495-500, 1990.

2. Bakker AR. Communication between hospital and radiology information systems and picture archiving and communication systems. In Osteaux M (ed.). Hospital integrated picture archiving and communication systems – a second generation PACS concept, Berlin: Springer, 55-97, 1992.

3. Baxter B.S., Hitchner L.E., Maguire G.Q., A standard format for digital image exchange, American Institute of Physicists in Medicine (AAPM) Report Number 10, New York: AAPM, 1982.

4. Deasy JO, Blanco AI, Clark VH. CERR: a computational environment for radiotherapy research, Medical Physics, **30**(5), 979-985, 2005

5. DICOM 3.0 Conformance of AccuTrans, DICOM implementation version 1.2, 2003.

6. ENDOBASE & OLYMPUS Medical System, DICOM conformance Statement, Revision 2.1.1, 2005.

7. Feingold E, Seshadri SB, Arenson RL., Folder management on a multimodality PACS display station, Proceedings SPIE, 1446:211-216, 1991.

8. Ferris, M., Lim. J., and. Shepard, D., An optimization approach for radiosurgery treatment planning, SIAM Journal on Optimization, 2003.

9. Ferris, M., Lim, J., and. Shepard, D., Radiosurgery optimization via nonlinear programming, Annals of Operations Research, 2003.

10. GE medical Systems, Conformance Statement for DICOM V3.0., 2003.

11. Hoehn H, Ratib O., Papyrus 3.0: the DICOM compatible file format, Geneva, Switzerland: Digital Imaging Unit, Center of Medical Informatics, University Hospital of Geneva, 1993.

12. Horiil S., F. Prior, W. Bidgood, *DICOM: An Introduction to the Standard*, Penn State University, the Radiology Department.

13. Hurson AR, Pakzad SH, Cheng J.-b., Object-oriented database management systems: Evolution and performance issues, IEEE Computer **26**(2), 48-60, 1993.

14. Horii SC, Seshadri SB., PACS clinical applications, Diagnostic Imaging, 15(9): PACS Supplement, 1993.

15. Intensity Modulated Radiation Therapy Collaborative Working Group, Intensity-modulated radiotherapy: Current status and issues of interest, International Journal of Radiation Oncology: Biology, Physics, **51**(4):880-914, 2001.

16. Jumppanen, S. Jaatinen, T. Toivanen, *DICOM XML*, Seminar in Medical Informatics, 2005.

17. Lead Tools Imaging developer SDKs, ***Overview: Compression/ JPEG Compression:*** *Sample DICOM Images using various compression methods.*
http://www.leadtools.com/SDK/Medical/DICOM/ltdc19.htm

18. Levine BA, Mun SK, Benson HR, Horii SC. Assessment of the integration of a HIS/RIS with a PACS. Proceedings SPIE, 1234:391-397, 1990.

19. Lim, G, Ferris, M., An optimization framework for conformal radiation treatment planning, accepted for publication, INFORMS Journal On Computing, 2005.

20. Medicor Imaging, MiPAC,
http://www.medicorimaging.com/Dental/DentalHome.

21. Prior, FW. Specifying DICOM compliance for modality interfaces. Report prepared under contract DAMD17-93-M-4464, U.S. Army Medical Research and Development Command.

22. Rorden, C., the DICOM standard,
http://www.psychology.nottingham.ac.uk/staff/cr1/dicom.html#header.

23. Reemtsen, R, Albert, M., Continuous Optimization of Beamlet Intensities for Photon and Proton Radiotherapy, Medical Physics and Biology, Vol. **59**, 2003.

24. Swiss Society of Radiobiology and Medical Physics,
http://www.sgsmp.ch/bull983b.htm

25. Varian Medical Systems, Inc., http://www.varian.com

26. Working groups of the DICOM Standards Committee, 2005.

Chapter 4

Reverse Engineering: A Review & Evaluation of Non-Contact Based Systems

Kevin D. Creehan, Ph.D.[1] and Bopaya Bidanda, Ph.D.[2]
[1]*Center for High Performance Manufacturing, Department of Industrial and Systems Engineering, Virginia Tech, Blacksburg, VA 24061, kcreehan@vt.edu,* [2]*Industrial Engineering Department, University of Pittsburgh, Pittsburgh, PA, bidanda@engr.pitt.edu*

Abstract

This chapter will define the concept of reverse engineering systems that are typically utilized in design and manufacturing environments. We will also develop a taxonomy of reverse engineering systems. Differences between contact and non-contact methods for reverse engineering will be detailed. Commonly used non-contact systems, including active and passive systems, will be detailed. Our focus will be on techniques such as laser scanning and 3D cameras.

Key words

Reverse engineering, non-contact, CAD, CNC, taxonomy, active techniques, passive techniques, laser scanning, STL format, CAM, photogrammetry, CCD, MRI, computed tomography, ultrasound scanning, medical image data file, ACR/NEMA, DICOM3, three-dimensional reconstruction, rapid prototyping

4.1 Introduction

Reverse engineering, widely noted as an effective cost saving tool, is a systematic approach used to analyze the dimensions, contours, and design of an existing device so that one may derive potential improvements to the device or perform competitive benchmarking to further understand the product. While reverse engineering is typically used as a manufacturing "aid," the resulting information contributes to product evolution, either at the subsystem, configuration, component, or parametric levels, that will occur through redesign processes.

The origins of reverse engineering are largely unknown. Rather than having been developed at a specific period or moment in time, the concept of reverse engineering arose in an evolutionary manner. It began as a way to "build a better mousetrap." As a design was developed, it would be critically assessed, examined, changed, and thus the design would inevitably be improved. Today reverse engineering is used not as a tool to be employed to an existing problem, but as practical methodology to new challenges of unique parts (with no design or developing computer drawings/molds of parts and tools where none exist).

Reverse engineering has been described as "a four-stage process in the development of technical data to support the efficient use of capital resources and to increase productivity"[6]. These stages are:

- Data evaluation – visual inspection, dimensional inspection, quality evaluation, possible failure analysis
- Data generation – engineering drawings, CAD models
- Design verification – prototyping, model testing, model failure analysis, quality assurance
- Design implementation – prototype delivery, project summaries, economic analysis, final implementation

Reverse engineering serves as a starting point in the product redesign process, during which the product is analyzed in terms of its functionality, physical principles, manufacture-ability, and assemble-ability, for the purpose of fully understanding every detail of the product. However, to begin the reverse engineering sequence, accurate geometric data from the surface of the existing part must be obtained.

4.2 Non-contact reverse engineering techniques

In order to accurately recreate the existing part, a computerized (CAD) model of the part's geometry must be developed in some manner. This CAD file provides the coordinates of multiple points on the product surface, which, in a manufacturing setting, is used to develop the drawing of the product for redesign or manufacturing. This data may then be analyzed and improved within the CAD program. The improved CAD design, along with other manufacturing information, is then used to create the manufacturing process plan and the Computer Numeric Control (CNC) tool path.

The software for non-contact reverse engineering systems greatly simplify the data gathering. Most parts are composed of standard shapes such as arcs, circles, spheres, and rods. A menu of standard shapes is generally available so that a user can specify one of these shapes, measure several points, then have the software complete the shape. Those shapes that are not standard can be digitized using one of several scan techniques in which the part is traced and the software is instructed to section the trace as parallel planes, radial sections, or concentric circles. The best approach to digitizing a complex shape is to divide it into simple zones and use the appropriate scanner technique.

Although not intended as a design medium, the software provides some ability to manipulate images by reversing, doubling, or repositioning them. If a part is symmetrical, half the part can be digitized then reversed in software and pasted to the first half. Once the basic image has been captured by the scanner, it can be transformed for its intended purposes by numerous software packages.

4.2.1 Reverse Engineering Taxonomy

There are several different methods that can be employed to collect surface data. These methods can be broadly classified into two categories: contact and non-contact methods. Contact methods, the more traditional manner of collecting data that has been utilized for several years, requires contact between the surface and a measuring device, usually a probe or stylus. Contact methods generally measure the surface of the object using a contact probe, a highly sensitive pressure-sensing device that is activated by any contact with an object. The linear distances from three axes to the position of the probe are ascertained, thus giving the x, y, and z coordinates of the surface. Conversely, non-contact methods typically use light, or laser beams, as the main tool for deducing surface information.

Non-contact techniques can be further classified into two additional categories: active and passive techniques[3].

Active techniques, organized into two subgroups, structured lighting and spot ranging, recover depth information after a light or ultrasonic source illuminates the object:

- Structured lighting methods are classified according to the pattern of light that is used to illuminate the object, such as single light beam, single stripe of light, and patterned lighting. Surface information is determined using triangulation procedures.
- Spot ranging techniques are generalized based on the source used, either optical-based or ultrasonic. These methods involve the projection of a beam onto the object surface and the inspection of the reflected beam using a sensor.

Both methods cast a beam onto the object surface and then inspect the reflected beam using a sensor that is placed coaxial to the source. The location of the source gives the two coordinate measurements of the surface point while the analysis of the reflected beam gives the third dimension. The third coordinate is determined either by calculating the phase difference between the incident and reflected light or by the time taken for the light to reflect back from the surface of the part. Typically, it is more accurate to use a method that calculates the phase difference of the incident light rather than one that calculates the time of reflection because this calculation of the time of refection relies on sensitive, accurate timing devices.

Passive techniques, work with ambient light, and are divided into three categories: stereo scanning, range from texture, and range from focus:

- Stereo scanning is accomplished by acquiring two or more images of the object from different perspectives. The corresponding surface points on the two images are identified, and a triangulation procedure is used to identify the location of points. As digital imaging and computational capabilities have improved with enhanced computer power, this process has become more popular.
- Range from texture processes are based on the fact that the further away from an object one is, the smoother its surface will appear to be. So if the texture is known, its distance from a known viewpoint can be estimated by inspecting the perceived texture at that distance. This method has limited accuracy and is not widely used in reverse engineering applications.
- Range from focus techniques are based on using the focal length of a lens to estimate the part's distance. This method is rarely used for reverse engineering purposes.

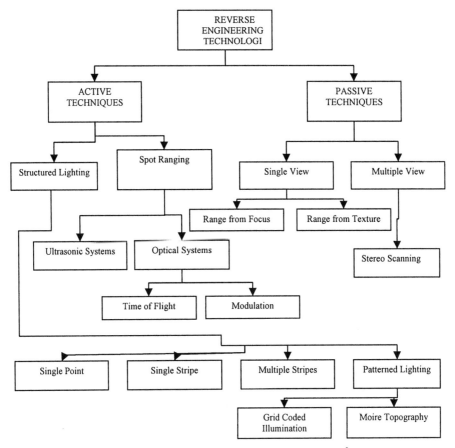

Figure 4-1. Reverse engineering taxonomy[2].

4.2.2 Active Technique – Laser Scanning

A well established active technology designed for accurate geometric data acquisition is laser scanning. Laser scanners, which utilize spot ranging techniques to acquire data, provide the ability to quickly acquire large amounts of geometric data. These systems measure points on the part surface by computing the position of a laser spot projected onto the surface of a part. The laser scanner casts a beam onto the object surface and then inspects the reflected beam using a sensor that is placed coaxial to the source. The location of the source gives the x and y coordinate measurements of the surface point while the analysis of the reflected beam by a triangulation procedure gives the third dimension.

A laser scanning system utilizes geometry acquisition technology that does not use imaging optics or sophisticated detector array processors. Rather, it is a linear device with none of the conventional focal length or image processing complexities inherent in other technologies.

Mechanically, the laser scanning systems may integrate rotation of the scanned part or translation/rotation of the source (laser). New technologies can now place the laser source on a probe capable of translating along all axes and rotating, giving it the ability to acquire data within deep concavities. This geometric data acquisition method provides high-resolution measurements across the entire work volume.

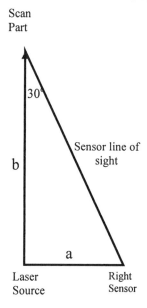

Figure 4-2. The triangle formed between the laser, the scanned part, and the sensors, which provides the geometry to obtain the depth coordinate

The top-down figure of the laser scanner (above) illustrates the functionality of these data acquisition devices. Composed of two triangulation sensors located on either side of a laser projector, all positioned on a single horizontal plane, the view line of each sensor is fixed to thirty degrees relative to the laser beam. This creates a 30°-60°-90° right triangle between the source, sensor, and part whose dimensions may be easily obtained with minimal information and simple geometric laws. To measure a surface point, the scanner is oriented so that the point's surface normal is parallel to the laser beam. Once the laser beam illuminates the desired spot on the piece, the sensors translate until the view line of one of the sensors intersects the point. Thus, because the horizontal distance (or "a" on the diagram) between the source and the sensor is known, and all three angles of the triangle are known, the depth measurement (or "b") is easily obtained. The location of the source provides the horizontal and vertical coordinates of the illuminated point, and the process outlined above provides the depth coordinate, thus creating the three-dimensional surface data.

The scanning process generally begins by scanning the bottom layer of the work volume and proceeds upward to each layer in user-defined intervals. Therefore, when the scanning process is complete, the data is organized as a collection of the cross-sectional layers. By coordinating the motion of the laser source, a computerized scanning algorithm may intelligently maneuver the laser about the object, exposing and recording all portions of the object's surface. The challenge remains in attempting to minimize shadowing effects that occur as a result of concavities in the part and to provide part orientations that optimize measurement.

Once the scanning procedure is complete, the resulting data must be analyzed to ensure its accuracy. Presently, the scanning technician performs this analysis manually, but algorithms have been developed to facilitate this process[5]. Upon assurance of accurate data, a triangulation procedure may be performed to integrate the three-dimensional data structure by joining the cross-sectional layers together. At this point, the file may be exported to many CAD data file types, including STL format.

The four-axis capability of the laser scanner allows for the acquisition of complex, three-dimensional surface data in one scanning session without having to reset and rescan the part. Laser scanning technologies are generally able to produce accurate geometric data repeatable to less than 0.001 inch. The data is highly compatible with most CAD software programs as well as rapid prototyping machines and Computer Aided Manufacturing (CAM) software, and can be exported as an ASCII file and translated into a STL file for rapid prototyping.

Figure 4-3. Three-dimensional laser scanners. The Roland LPX-250 (left) incorporates rotary and plane scanning and the handheld Modelmaker Z (left) manufactured by 3DScanners, Ltd. is ideal for parts with limited access.

4.2.3 Passive Technique – Three-Dimensional Photogrammetry

Stereo scanning, a passive reverse engineering technique, is the process of acquiring geometric data of an object by integrating multiple photographs of the object from different perspectives. The primary challenge of this method is the identification of the matching image pixels that correspond to the same point on the object. Traditionally, this process has been used from large distances, for example to develop terrain maps. This is accomplished by taking multiple photographs of a terrain while flying in an airplane.

Close range photogrammetry has taken hold in recent years as processing power and digital imaging has improved. In this method, images are taken from positions all around the object. Mathematical models have been devised to produce dimensional coordinates of discrete points on the object. A significant challenge in this process is the conversion between coordinates in systems having different origins, orientations, and scales. The fundamental principle used by photogrammetry is triangulation. By taking photographs from at least two different locations, lines of sight can be developed from each camera to points on the object. These lines of sight are mathematically intersected to produce the three-dimensional coordinates of the points of interest. These coordinates are then estimated using a least squares approximation, with a large number of degrees of freedom[1].

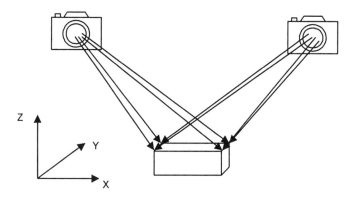

Figure 4-4. The triangles formed between the multiple image perspectives

A two-dimensional image may be regarded as a slice through the projections of object points in space. Because all space points along a projection line will appear at the same point on the image plane, a unique determination of the three-dimensional position is not possible. The introduction of additional images allows a unique description of the object point as the intersection of two or more image planes. This unique three-dimensional description requires knowledge of the orientation of the cameras with respect to a common frame of reference, which requires camera calibration. This determines the set of parameters that relate the two-dimensional image position of an object point to its corresponding three-dimensional position in the reference coordinate system. This mapping is then used to obtain the three-dimensional coordinates of points occurring on two or more two-dimensional images[17].

A stereo photogrammetric platform typically consists of two or more cameras which capture images of the subject at the same time, or one or more cameras capturing images sequentially with the subject fixed. The scanned part is placed with a calibration object or in a fixed location where a calibration object has been captured. The success of such methods often depends on a consistent lighting environment. For purposes of surface reconstruction, photographs are usually taken under controlled conditions, and preliminary estimations of feature positions are possible, eliminating some of the preprocessing required in other reverse engineering technologies and applications[17].

A charge-coupled device (CCD) is an integrated circuit containing an array of linked or coupled, capacitors. Under the control of an external circuit, each capacitor can transfer its electric charge to one or other of its neighbors. CCD cameras are used in digital photography and astronomy to obtain three-dimensional data quickly. These cameras have the following advantages in addition to the advantages of all non-contact systems:

- very fast scanning time
- integration of multiple cameras is possible, allowing for stereo scanning
- not sensitive to the color of scanned parts
- able to scan very small areas[16].

However, these systems have the following notable disadvantages:

- very high price
- accuracy decreases as the camera distance increases
- scanning angle is the same regardless of the shape of the part's surface, so holes and steep angles often produce poor data
- the presence of dust, oil, and water can produce disfigured data[16].

Figure 4-5. Samsung SCC-B2305 CCD camera.

4.2.4 Medical Imaging

The rapidly growing field of medical imaging has become a cornerstone of the health care industry today. The advancement of digital technology, specifically mass data storage and expeditious data acquisition has led to explosive growth in the industry through the development of several new imaging technologies[10]. However, the development of these new techniques has not only improved the quality of health care; it has also contributed to its rising cost. Imaging methods such as Magnetic Resonance Imaging or Computed Tomography can now acquire accurate three-dimensional data with excellent precision, but the cost of these imaging methods can be prohibitive. Costs vary based on the particular protocol used and the part of the body being scanned, as well as the availability of and local demand for the technology.

Collecting extremely accurate geometric data from human structures is generally difficult due to their complex shapes and the flexibility of soft tissues. In addition, collecting geometric data from organs within a living subject requires a non-invasive imaging technique. Thus, these imaging methods, having these complex abilities, are frequently used for diagnosis of various clinical symptoms despite their considerable expense.

4.2.4.1 Magnetic Resonance Imaging

Magnetic Resonance Imaging (MRI) is a medical diagnostic technique that creates images of the body using the principles of nuclear magnetic resonance. A versatile, powerful, and sensitive tool, MRI can generate thin-section images of any part of the body—including organs, bones, and ligaments—from any angle and direction, without surgical invasion and in a relatively short period of time.

The principles of MRI take advantage of the random distribution of protons, the nuclei of abundant hydrogen atoms, which possess fundamental magnetic properties[10]. Once the patient is placed in the cylindrical magnet, the diagnostic process follows three basic steps. First, MRI creates a steady state within the body by placing the body in a steady magnetic field that can be over 30,000 times stronger than the earth's magnetic field. Then the MRI scanner stimulates the body with radio waves to change the steady-state orientation of the protons. After terminating the radio signal, a short-wave radio antenna detects, at a pre-selected frequency, the electromagnetic transmissions emitted by the hydrogen nuclei. The transmitted signal is used to construct internal images of the body using principles similar to those developed by computed tomography, or CT scanners.

MRI utilizes nuclear magnetic resonance, a process that utilizes the absorption and emission of energy by atomic nuclei in the presence of an externally applied magnetic field. The patient is placed in the MRI scanner and energy is applied in the form of radiofrequency pulses that are designed to match the resonant frequency of hydrogen atoms, which absorb these energy pulses. Hydrogen proton nuclei are used because they are the most abundant nuclei in the body, with large concentration in water and lipid molecules. After each radiofrequency pulse has been applied, the hydrogen atoms re-emit this energy as a magnetic resonance signal that induces a small voltage in a receiver coil. The hydrogen atoms then return to a relaxed state[11].

Due to the strong magnetic field, metallic objects are potentially dangerous within the MRI. Orthopaedic implants including prosthetic joints are acceptable, but they significantly degrade image quality. The strength of the magnet correlates with image quality, although very high field strengths may also lead to increased degradation. Low field strength magnets degrade image quality and lengthen examination time, thus further degrading the image by patient movement[11].

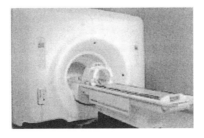

Figure 4-6. General Electric 3.0 Tesla Signa VH MRI scanner is the first FDA-cleared 3.0T system in the industry for whole-body imaging.

4.2.4.2 Computed Tomography

The Computed Tomography (CT) scanner, an intricate extension of the conventional X-ray device, offers clear views of any part of the anatomy, including soft organ tissues. Used without the need for injected dyes, the full body scanner rotates 180° around a patient's body, sending out a thin X-ray beam at hundreds of different points[10]. Detectors, usually solid-state crystal or Xenon gas, positioned at the opposite points of the beam detect and record the absorption rates of the varying densities of tissue and bone. The density of the scanned object determines the quantity of the emitted X-ray beam that passes through the object, and thus the intensity of the grayscale image at that location. These data are then relayed to a computer that converts the information into a picture.

The fundamental principle of CT is to irradiate a slice of tissue from multiple angles measuring the output from the opposite side. Different tissues have different densities, and thus attenuate the x-ray beams to varying amounts. Fewer X-rays will reach the detectors when traveling through denser tissue. Multiple measurements are taken, and using Fast Fourier Transformations, an image of the different density tissues within the slice can be produced by assigning a grayscale to the resulting values[12].

The first CT scanners were very slow and required a water bath around the area of interest, significantly limiting the range of applications. In this process, a single x-ray source and detector were rotated around a fixed point obtaining data from a single location for processing before moving onto the next point. Advances in this technique and its applicability came with the development of spiral scanning. In this scanning method, the table moves during the scan resulting in a spiral data acquisition through the volume of interest. The spacing between the spirals directly affects the scanning detail, as small layer distances produce more detailed scans. However, small layer thicknesses limit the scanning volume[12].

Subsequent technology improvements have led to the development of multi-slice CT scanning. The use of multiple detectors simultaneously increases the amount of data that can be acquired in a single rotation. With this method, it is now possible to get far more accurate representations of the anatomy very quickly. Full body scans are now possible in mere minutes[12].

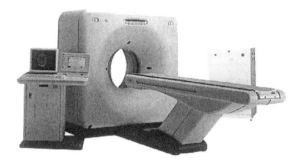

Figure 4-7. General Electric Sytec 1800i CT scanner

4.2.4.3 Ultrasound Scanning

Ultrasound is a sound wave having a frequency higher than the upper limit of the human audible range, 20 kHz[10]. Ultrasound Scanning (US) is a non-invasive diagnostic medical imaging technology that uses these high frequency sound waves to form an image of body tissues that can be viewed for medical diagnosis. The sound waves are recorded and can be displayed as a real-time visual image or simply as a still image. As the sound waves pass through the body, echoes are produced that can be used to identify how far away an object is, its size, and its uniformity.

The ultrasound transducer functions as both a generator of sound and a detector. When the transducer is pressed against the skin, it directs the inaudible, high frequency sound waves into the body. As the sound echoes from the body's fluids and tissues, the transducer records tiny changes in the pitch and direction of the sound. These echoes are instantly measured and interpreted by a computer, which in turn creates a real-time picture on the monitor.

Ultrasound scanning is an operator-dependent technique that requires experience. Yet, ultrasound has many advantages over other cross-sectional imaging techniques with the best available soft tissue resolution, lack of ionizing radiation and the potential for real-time guided intervention[13].

4.2.4.4 Medical Image Data File

The data formats for medical imaging devices, specifically MRI images and CT scans, are extremely similar. Standard formats have been developed and are adhered to in varying degrees. However, some imaging equipment proprietors have developed their own proprietary formats that do not resemble the standard formats. Three types of information are generally present in medical imaging data files:

- Image data, which may be unmodified or compressed
- Patient identification and demographics
- Technique information about the exam, series, and slice/image

Although extracting the image information is usually straightforward, the extraction of descriptive information such as geometric details in order to combine images into three-dimensional data sets is more difficult and requires a deep understanding of how the files are constructed.

There are three basic families of formats in use:

- Fixed format, where layout is identical in each file
- Block format, where the header contains pointers to information
- Tag based format, where each item contains its own length

The block format is a popular type used, though in most cases the header contains only a limited number of pointers to large blocks and the blocks are almost always in the same place and a constant length. Thus, a user can succeed by assuming a fixed format if the specifics of the layout are unknown.

The American College of Radiologists (ACR) and the National Electrical Manufacturers Association (NEMA) developed the first standard in 1985, referred to as ACR/NEMA version 1.0, to facilitate multi-vendor connectivity. After several revisions, the standard was re-released as version 2.0 in 1988. The most recent version was released as Standards Publication PS3 but is commonly referred to as DICOM 3 (Digital Imaging Communications in Medicine)[14].

In the ACR/NEMA format, the presentation of the data and the application layers are described in a message format, which, along with the ACR/NEMA data dictionary and extension mechanisms, has been adopted by many proprietary formats and other standards. A message consists of a series of data elements, each of which contains a piece of information. Each element may occur only once in a message. Even numbered groups describe elements defined by the standard. Odd numbered groups are available for use by vendors or users, but must conform to the same structure as standard elements.

An unwritten standard has developed among the vendors of the equipment to almost universally put the image data at the end of the file. Many vendors, including Siemens, Philips, General Electric, Toshiba, Hitachi, Imatron, and many others, have developed their own proprietary formats. Although they differ in varying degrees from the standard formats, these proprietary formats generally utilize many of the same concepts as the standard formats.

4.2.5 Three-Dimensional Reconstruction

A problem that continues to be studied is the three-dimensional reconstruction of parts from parallel cross sections, such as those produced by MRI, CT, and laser scanning systems. In medical industries, three-dimensional models are used for surgery planning, prosthesis fabrication, radiation therapy planning, and volumetric measurement. In manufacturing industries, three-dimensional models are used to modify part design and to develop prototypes.

The utilization of computer graphics technologies in order to visualize, analyze and reproduce the human body has become crucial to the present medical field. Two-dimensional data generated by medical imaging systems such as CT, MRI, and US are often not comprehensive enough for the surgeons to analyze the conditions of their patients practically[15]. Three-dimensional reconstruction techniques rebuild a computer model of the object that has been reduced to slices by the imaging method. The advantage of a three-dimensional reconstruction is that a global view of the structure is obtained, eliminating the need to reconstruct the object from multiple cross-sections by inaccurate estimation. Three-dimensional models can provide the following benefits in medical industries:

- Quantitative morphological description of the patient's anatomy
- Pre-operative planning and simulation of surgical methods
- Biomechanical modeling of joints
- Intra-operative registration of patient with 3-D data set and precise navigation of surgical tools, in conjunction with computer assisted surgery
- Development of a knowledge base
- Manufacturing of precise physical models and implants

The need for accurate three-dimensional reconstruction has inspired its own rapidly growing field of study within the computer science and engineering fields. Despite the advantages it provides, the representation of a three-dimensional volume on a two-dimensional monitor does not necessarily provide the necessary information to understand the three-dimensional geometry of the part. In order to extract three-dimensional information from the picture, users must learn to interpret on-screen displays. There are many visualization issues that are not resolved by computer models:

- Two-dimensional screen displays do not always provide an intuitive representation of 3-D geometry
- Unusual or deformed geometry may be difficult to observe on a monitor
- The integration of multiple assembled parts is difficult to visualize
- The integration of multiple assembled parts is difficult to visualize

As such, significant research has been performed to develop cohesiveness between the medical technologies, reverse engineering technologies, and rapid prototyping technologies, so much that conferences are now being held solely to further research interests in this topic. Many software programs have successfully bridged the gap from medical imaging data to rapid prototyping fabrication.

4.3 Applications

Applications of non-contact reverse engineering in manufacturing are widespread. There are many instances where an existing part or prototype requires reproduction. Often a new product does not start from a CAD model but rather a prototype is built first. Upon completion, the measurements of the new prototype are taken. Also, design changes are often required to an existing product for which CAD data does not exist. Numerous additional applications of reverse engineering in manufacturing exist and include: the design of large equipment whose measurements can not be taken using metrology, replacement of worn or broken parts for which CAD data is unavailable, and inspection of a produced part compared to its original CAD design.

Applications of the reverse engineering process in the biomedical community are similarly popular, although performed with different methods. For years, the medical community has used several non-invasive imaging devices to produce two-dimensional cross-section images of patients. As computer technology has grown, software products have been developed that extract three-dimensional images from this data. At this point however, with only a few exceptions such as surgery planning and prosthesis milling, the similarities to a manufacturing-style reverse engineering and redesign methodology typically end. Comparing the current methods used by the medical community to the four reverse engineering and redesign stages described by Ingle[6], one can see a relationship between the manufacturing process Ingle described and the processes the medical community applies.

Rapid Prototyping: Theory and Practice 103

- Data evaluation – patient examination, preliminary diagnosis, etc.
- Data generation – development of medical image (MRI, CT, US, etc.)
- Design verification – 3-D computerized reconstruction, physical model generation
- Design implementation – final diagnosis, patient treatment, etc.

There are many instances in which one-of-a-kind parts such as prototypes or custom-built parts need to be produced. Moreover, the design of the existing manufactured parts may require periodic modifications to update and improve. CAD models of these parts are often not available. However, the creation of a CAD model is desirable and often necessary under the following circumstances:

- The design process of a new product does not always start from a CAD model. A prototype is often built first. Once the design is approved, measurements are made. The extracted data are then manually entered into the computer. This process has two primary disadvantages: it is time-consuming and a potential source of measurement errors.
- In some cases, the design of an existing product must be modified. The modification process and design improvements are best performed on a CAD model. However, CAD models for many existing products are unavailable. Part image reconstruction systems can play an important role in reducing redesign time.
- Precise measurements of large parts are often not available. Reverse engineering-based part image reconstruction systems can help by mapping the part surface in the form of a CAD model. This CAD model can then be scaled and modified as needed.
- When a one-of part breaks or is worn and the engineering drawing is no longer available, part image reconstruction systems can be used to create a new CAD model. The CAD model can now be used to manufacture a replica of the worn or broken part.
- When a part is compared with its existing CAD model, a reverse engineering system can acquire the actual map of the part surface, and deviation, if any, can be identified[3].
- When a part is compared with its existing CAD model, a reverse engineering system can acquire the actual map of the part surface, and deviation, if any, can be identified[3].

4.4 Relationship to Rapid Prototyping

Rapid prototyping provides the means to dramatically reduce the lead-time required to produce a semi-functional physical prototype. The ability to produce these models in a matter of hours is possible due to techniques that implement an additive fabrication process. While most traditional manufacturing techniques use subtractive or formative processes, RP techniques build the model, one layer at a time, from bottom to top. This additive nature provides cohesiveness with the data types created by reverse engineering systems, as they are created in sliced intervals as well. As a result, any model can be quickly cloned. Once the data has been successfully acquired by the reverse engineering system, limitless design alternatives, achieved through modification or scaling of the newly-acquired CAD data, may be fabricated into physical prototypes to achieve any of the aforementioned manufacturing objectives.

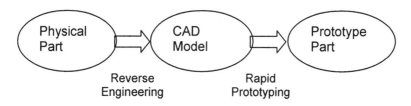

Figure 4-8. Relationship between reverse engineering and rapid prototyping

The practical integration of non-contact reverse engineering and rapid prototyping technologies is becoming more prevalent each year. Fields such as engineering, automotive, aerospace, medical, entertainment, electronics, and consumer goods are taking advantage of these easily-integrated technologies to develop creative solutions to design problems. The applications, when combined with creativity, are almost endless. Following are some common applications that illustrate use of non-contact reverse engineering and rapid prototyping technologies.

Application 1 – Medical Industry

Rapid prototyping, being a fast accurate method of building mechanical prototypes, offers great potential in medicine when utilizing geometrical information of anatomical structures from medical data. The tangible 3D information of a solid model combined with the extra information from selective coloring is a definitive diagnostic and preoperative planning tool.

While this process is still relatively new and unknown to many professionals working in the medical sector, the number of practical applications continues to increase as success stories continue to be shared.

Rapid prototyping represents a breakthrough in human modeling. Medical images can be used from any CT or MRI scanner, and format recognition for most scanner formats is available for processing the data. Precise medical models deliver visual and tactile information for diagnostic, therapeutic, and didactic purposes. The communication of pathology among the medical team and between the surgeon and the patient is facilitated. These models are used for the planning, and even rehearsing, of complex surgery, and medical models can act as a master or negative for the design of customized implants.

Application 2 – Automotive

The automotive industry has developed interesting applications of non-contact reverse engineering technologies for tool making and product testing. Using a physical model, dimensions can be taken to create everything from molds to fixtures for robotic welders. The same process is used to adjust tooling to dial in specifications. For some complex automotive assemblies, fabricators have cut almost a year off the time to qualify parts.

Application 3 – Army Mobile Parts Hospital

Long lead times and the high costs of procurement and inventory for replacement parts have reduced equipment readiness rates for the Army. The Army's Mobile Parts Hospital project exists to offset this imbalance by producing spare parts through rapid prototyping near the point of need. Laser range finder and high-resolution single-line CCD cameras can be used to capture high geometry data that are needed for rapid prototyping. This technology can also be applied to the inspection of the underside of vehicles for explosives and other dangerous elements, a topic of special interest in recent years.

References

1. K.B. Atkinson, Close Range Photogrammetry and Machine Vision, 2nd Ed., Whittles Publishing (2001).
2. B. Bidanda, S. Motavelli, and K. Harding, Reverse Engineering: An Evaluation of Prospective Non-Contact Technologies and Applications

in Manufacturing Systems, International Journal of Computer Integrated Manufacturing, Vol. 4, No. 3, pg. 145-156 (1991).
3. B. Bidanda, V. Narayanan, and Richard Billo, Handbook of Design, Manufacturing, and Automation, Ch. 48, Reverse Engineering and Rapid Prototyping, John Wiley & Sons, Inc. New York, (1994).
4. M. Butson and W.H. El Maraghy, Evaluation of Reverse Engineering Techniques, Autofact '96: Rapid Design and Manufacturing Conference Proceedings, Detroit, pg. 48-64 (1996).
5. K.D. Creehan, Computer-Aided Reverse Engineering of Human Tissues and Structures, Ph.D. Dissertation, Department of Industrial Engineering, School of Engineering, University of Pittsburgh (2001).
6. K.A. Ingle, Reverse Engineering, McGraw-Hill, Inc. New York (1994).
7. E. Mikhail, J. Bethel, and J. McGlone, Introduction to Modern Photogrammetry, John Wiley & Sons (2001).
8. K. Otto and K. Wood, Product Design: Techniques in Reverse Engineering and New Product Development, Prentice Hall, (2000).
9. K. Otto and K. Wood, Product Evolution: A Reverse Engineering and Redesign Methodology, Research in Engineering Design 10(4): 226-243 (1998).
10. Shung, K. Kirk, et al., Principles of Medical Imaging, Academic Press, Inc. New York (1993).
11. S. McKie and J. Brittenden, Basic Science: Magnetic Resonance Imaging, Current Orthopaedics, Vol. 19, pg. 13-19, (2005).
12. D. Barron, Basic Science: Computed Tomography, Current Orthopaedics, Vol. 19, pg. 20-26 (2005).
13. K Colquhoun, et al., Basic Science: Ultrasound, Current Orthopaedics, Vol. 19, pg. 27-33 (2005).
14. Digital Imaging and Communications in Medicine (DICOM) Standard, National Electrical Manufacturers Association, http://medical.nema.org/dicom/2004.html (2004)
15. B. Geiger, and M. Ioannides, Reverse Engineering and Rapid Prototyping Techniques in Medicine, International Conference on Medical Physics and Biomedical Engineering, University of Cyprus, Vol. 1, pg. 48-52 (1994).
16. M Sokovic and J. Kopac, RE (reverse engineering) as necessary phase by rapid product development, Journal of Materials Processing Technology, article in press (2005).
17. T. Douglas, Image processing for craniofacial landmark identification and measurement: a review of photogrammetry and cephalometry, Computerized Medical Imaging and Graphics, Vol. 28, pg. 401-409 (2004).

Chapter 5

Reverse Engineering: A Review & Evaluation of Contact Based Systems

Salil Desai[1] and Bopaya Bidanda[2]
[1]*Department of Industrial & Systems Engineering, NC A&T State University, Greensboro, NC 27411;* [2]*Department of Industrial Engineering, University of Pittsburgh, Pittsburgh, PA 15261*

Abstract

This chapter focuses on contact based reverse engineering systems. The major technique that we describe here is the utilization of co-ordinate measuring machines (CMMs). Different types of CMMs available are detailed. In addition the performance parameters of these systems are discussed. This chapter also explains the integration of data acquired from reverse engineering techniques with other design and manufacturing related software systems. This chapter concludes with a brief description of the state-of-the-art CMM technologies.

Key words

AFM; computer aided design; computerized numerical control; contact based systems; coordinate measuring machines; data collection; digitization; haptic volume sculpting; IGES; MEMS; Nano CMM; performance parameters; point cloud; point preprocessing; probe accuracy; reverse engineering; rigid body errors; scanning speed; STM; structural deformations; surface fitting.

5.1 Introduction

Reverse Engineering (RE) is the process of generating a Computer Aided Design (CAD) model from an existing physical part[1]. It enables the reconstruction of an object by capturing the component's physical dimensions and geometrical features. Essentially, it is a converse product design approach where the designer begins with the product and works through the design process in an opposite sequence to arrive at product specifications such as dimensions and form. This enables the designer to mentally simulate design ideas that occur during the design of the original product. Reverse engineering is accomplished in three steps[2]:

- Part digitizing,
- Feature extraction or data segmentation, and
- CAD or part modeling

As described in an earlier chapter, reverse engineering methodologies can be broadly classified as (a) non-contact based and (b) contact based systems. Non-contact based measurement systems include laser scanning though structured lighting[3], the Moire method[4], and phase shift interferometry[5] techniques. In spite of achieving high measuring accuracy, they are generally limited to conditions such as sensitivity to vibration, improper scanning due to reflective surfaces and the inability to react to drastic change in surface curvature[6] such as concavities and hidden surfaces. Contact based methods, the more traditional manner of collecting data that has been utilized for several years, requires contact between the surface and a measuring device, usually with a probe or stylus. Contact based measurements are usually more reliable, but more time consuming.

5.1.1 Need for Reverse Engineering

In today's competitive market lead time compression in a product development cycle with increased product variety is the essence of survival for an organization. Reverse engineering methodologies play a critical role in assisting designers and manufacturers in shortening time to market for new products. Some reasons for reverse engineering of component or product assemblies are as follows:

- Manufacture of service components after discontinuation of the product line
- Redesign of an existing design which lacks adequate product data documentation
- Corrupt data file or loss of CAD design for a product
- Competitive benchmarking of product components
- Generation of cheaper alternate products as a substitute for monopoly products

5.2 Contact Based Reverse Engineering Systems

Currently, the best results in terms of accuracy and quality of surface finish are obtained using contact reverse engineering systems. Contact systems have several fundamental advantages over non-contact systems[7]. These include:
- Treatment of surfaces to prevent reflections is not required
- Vertical faces can be accurately scanned
- Data density is not fixed and is automatically controlled by the shape of the component
- Time-consuming manual editing of the data to remove stray points is not required
- Post processing for cutting can be faster as surface offsetting may not be required
- Very tiny details can be accurately replicated

Contact scanners are stand-alone scanning systems with these advantages and are evaluated based on additional qualities that separate themselves from other systems of this type. For example, those systems that require low probing forces, allows the user provides the ability to accurately scan delicate materials, and are able to utilize extremely small styli (0.3mm), which allows the scanning of very fine detail, are the most commonly used systems[6]. Other advantages of using the contact scanning systems include quiet and clean operation and the magnetic breakaway ability of the stylus, which provides damage protection for both the work piece and stylus in a crash situation caused by an improper set up by the user[6].

The ability to define planes and align axes in conjunction with a full range of datum functions makes complex fixtures unnecessary and allow for rapid job set up. The many options for data capture allow almost any free form shape, no matter how detailed, to be scanned[8]. Contact scanning systems combine output from their scanning probes and axes positions using a special scan control card for data capture. The accompanying software calculates the surface coordinate data point and a new target position to which the machine should move by analyzing the force placed upon the stylus. The stylus then moves to the new position automatically[7].

Scanning software packages are used to digitize original models on both computerized numerical control (CNC) machine tools and coordinate measuring machines (CMM) to produce mold and die cavities, CAM profiles, and templates. Data can be captured using touch-trigger or analog probes. Once the data has been captured, model variants can be produced by mirroring, scaling, rotation, translation and male/female inversion[6].

Some systems maintain a contact pressure of only few grams and are able to translate a model into pure data that can then be processed for use by either manufacturing or engineering purposes. Once the scan is complete, the software enables the user to manipulate data and then create an NC program or export the data to any of several CAD output file types[7].

The contact systems' touch-trigger probes can be used to digitize components when fitted to either a CNC machine tool or coordinate measuring machine. The analog probes operate at very high scanning speeds, with greater accuracy and lower contact force than existing tracer probes[6]. This allows better machine finishing of parts, and the ability to scan relatively soft materials. The scanning probes have been developed for the intricate scanning of objects such as coins, watermark dies, and jewelry, where fine detail requires the use of very low contact force[6]. The contact scanning technology's flexible scanning solutions can provide the answers to most scanning problems

5.3 Coordinate Measuring Machine (CMM)

Contact based methods of reverse engineering have been available for nearly forty years[9]. The first (and still the most popular) method of reverse engineering, to be introduced was the coordinate measuring machine (CMM). A Coordinate Measuring Machine gives physical representation of a three-dimensional Cartesian coordinate system[10]. The CMM measures the surface of the object using a contact probe, a highly sensitive pressure-sensing device that is activated by any contact with an object. The linear distances from three axes to the position of the probe are ascertained, thus giving the x, y, and z coordinates of the surface.

Other contact methods of reverse engineering include electromagnetic digitizing and sonic digitizing[11]. Electromagnetic digitizers determine the surface data of non-metallic objects (placed in a magnetic field) by tracing a hand-held stylus containing a magnetic field sensor across the surface of the object. The magnetic field sensor, in conjunction with an electronic unit, detects the position and orientation of the stylus. The sonic digitizer uses sound waves to calculate the position of a point relative to a reference point. Again, using a hand-held stylus tracing the surface, an ultrasonic impulse is emitted by the stylus and is detected by four microphones. The times taken to reach each of the four microphones are recorded and the computer calculates the x, y, and z coordinates from these time differences[11].

5.3.1 Types of CMM Configurations

The type of configuration for a CMM determines its measuring parameters such as accuracy, flexibility, time or throughput of measuring process, maximum measurable workpiece dimension and cost. Most of these configurations can be controlled manually by an operator or by a program. An operator driven CMM is called a Manual CMM and the one that runs by a program is called a CNC CMM. CNC stands for Computer Numerical Control. This section details the CMM configuration with their relative advantages and disadvantages and applications.

5.3.1.1 Bridge Type

In bridge style machines, the arm is suspended vertically from a horizontal beam that is supported by two vertical posts in a bridge arrangement. The machine x-axis carries the bridge, which spans the object to be measured. There can be inconsistencies in the motion of the vertical posts while spanning the table length, causing the bridge to twist or yaw. This error can be corrected using a positive position feedback control system or a central drive mechanism which moves both posts at the same time. Given the rugged construction of this machine, it has higher natural frequencies which improve the dynamic response of the machine as compared to the overhanging cantilever type CMMs. This type of CMM can have a smaller footprint which is suitable for clear room or design laboratory type of facilities. Figure 5-1 shows examples of bridge type CMM's.

5.3.1.1.1 Applications[12]
- Mechanical parts inspection
- Digitalization and inspection of complex mechanical components (gears, cams, airfoils/turbine blades)

- Free form surfaces inspection (dies, models, sheet metal, plastic, moulds)
- Point to point inspection
- Continuous scanning inspection

5.3.1.2 Gantry Type

This class of machines is used for large part sizes which can span 4 meters or more. Gantry style machines employ a frame structure raised on side supports so as to span over the object to be measured or scanned. A horizontal beam traverses the length of the measured object. It is powered with dual drives so as to minimize the yaw or twisting of the side supports during traverse. A measuring arm is mounted on this horizontal beam that moves along the width of the object being measured. Gantry machines have a rugged construction as compared with other CMMs. This is essential to offset the deformation caused by twisting and the weight of the measured part on the foundation. This type of machine has the best measuring volume to overall dimension ratio within CMMs. Additional precision can be enhanced using thermal compensation and combination of air bearings and high accuracy linear guide-ways. Figure 5-2 shows examples of gantry style CMMs.

Figure 5-1. Examples of Bridge type CMMs[12]

5.3.1.2.1 Applications
- Inspection of large components (such as pipes, pressure vessels, automobile frames)
- Measurement of gages, and fixturing systems for heavy and large complex parts
- Shop-floor inspection equipment with high operational safety

Figure 5-2. Examples of Gantry type CMMs[12]

5.3.1.3 Cantilever Type

In cantilever style machines, a vertical arm is supported by a cantilevered support structure. This type of open configuration allows for easy operator access to the object being measured. Heavy parts can be measured by placing them on the fixed table. However, due to the overhanging cantilever structure it has a lower system natural frequency affecting the speed of measurement. This type of system is suitable for longitudinal parts that fit along the length of the table and have smaller dimension in the other two axes. Examples of Cantilever type CMMs are shown in Figure 5-3.

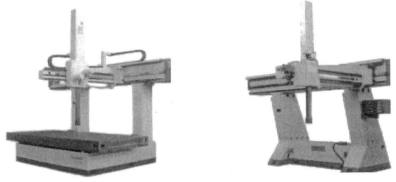

Figure 5-3. Examples of Cantilever type CMMs[12]

5.3.1.3.1 Applications

- Marking-out on models, casts and sheets
- Light milling operations
- Copying of free form surfaces

5.3.1.4 Horizontal Arm Type

Horizontal arm machines are widely used in the automotive industry. In this configuration the arm that supports the measuring probe is horizontally cantilevered from a movable vertical support. As a result, this style is sometimes referred to as cantilever design. It is also available in dual arm configuration as shown in Figure 5-4. The overhanging arms limit the dynamic stiffness of the machine affecting speeds of measurements. However, this error can be compensated with software correction.

Figure 5-4. Dual Arm Type Horizontal CMM[13]

5.3.1.4.1 Applications
- Inspection of large dimension components in one axis
- Measurement of prismatic elements within auto subassemblies
- Verification of free form body contour (automobile styling and aircraft aerofoil shape)

5.3.1.5 Articulated Arm Type

An articulated arm configuration is used for portable, or tripod mounted style machines. The articulating arm allows the probe to be placed in different directions with respect to the object being measured. These systems contain a series of counterbalanced six-degrees-of-freedom linkage arms as shown in Figure 5-5. Each of the arms is provided with precision rotary transducers that encode the rotary motion of the linkages and calculate coordinates in 3D space. The measuring envelope of this type of system is spherical, enabling measurement of hard to reach locations within components. Accuracy of measurements is largely affected by operator skill and is lower as compared to bridge style systems. These types of systems are manufactured with light weight alloy for high rigidity and low weight. Given the portable nature of equipment, they are configured to operate at temperature ranges as high as 50°C. Arm type CMMs are versatile systems with a wide range of accessories for on-site measurement tasks.

Figure 5-5. Articulated Arm Type CMM[12]

5.3.1.5.1 Applications

- Suitable for field use for wide range of applications
- Measurement of subassemblies within very large systems (engine component within aircraft)

- On-the-fly inspection of basic dimensions on hard to reach features on the part
- Continuous measurement of free form surfaces (auto styling, aero-wing aerofoil contour, etc.)

5.3.2 Specifications of Coordinate Measuring Machines

Coordinate Measuring Machines (CMM) are electro-mechanical systems designed to traverse a measuring probe across the surface of a workpiece to determine its point coordinates. Essentially, the CMM is comprised of four critical components: the machine structure including bed, the measuring probe, the control system, and the integration software. Machines are available in a wide range of sizes and designs with a variety of probe technologies. Important specifications for coordinate measuring machines are the measuring lengths along the x, y and z axes as well as resolution and work piece weight. The x, y and z axes measuring lengths are the total travel, or measuring length that can be performed in the *x, y, and z* directions respectively. It is important to differentiate between measuring length and measuring capacity of a CMM. The earlier is total travel length, however, the later is the maximum size of the object in the x, y or z-direction that the machine can accommodate. The resolution is the least increment of a measuring device; on digital instruments, it is the least significant bit (Reference: ANSI B-89.1.12). The workpiece weight is the mass of the workpiece being measured.

5.3.2.1 Control types

Coordinate machines may have manual control, CNC control or PC control. Manual control implies that machine positioning is operator controlled. A Computer Numerical Control (CNC) may also control machine positioning. Personal computers (PCs) also control machine positioning in some coordinate measuring machines.

5.3.2.2 Mounting options

Mounting options include benchtop, free standing, handheld and portable. Manufacturers may use these terms interchangeably. Probe systems for CMMs can be touch probe or discrete point, laser triangulation, camera or still, and video camera. A multi-sensor <u>coordinate measuring machine</u> has capabilities to mount more than one sensor, camera, or probe at a time.

5.4 CMM Measurement process

A typical CMM measuring system includes a CNC machine, a scanning probe, 3-axis simultaneous motion drivers and counters, and a personal computer. Figure 5-6 shows the schematic diagram of a typical measuring system[6]. The measured object is positioned on top of the XY stage. The scanning probe is mounted in the vertical direction (Z stage) of the CNC machine. The probe stylus picks up the signal when contacted with the object or workpiece and transfers it to the digital signal processor and counter. Each of the three CNC stages can be moved simultaneously in the X, Y and Z axes simultaneously. When the probe is traversed along the workpiece surface, there exists a settling time for signal pickup, which is influenced by the probe's contact pressure. The linear pattern traversed by the probe during scanning is shown in Figure 5-7[6].

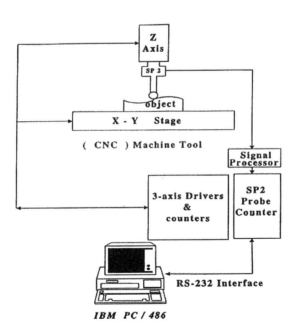

Figure 5-6. Schematic Diagram for a CMM[6]

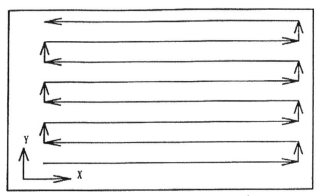

Figure 5-7. Scanning Probe path[6]

5.4.1 Data Collection Procedure for CMM

The touch trigger probe is traversed along the dimension to be measured. Based on the geometry of the object surface, the measured points usually form irregular grids or meshes as shown in Figure 5-8[6]. In order to determine the surface normal along the surface under inspection, these points need to be re-sampled or reassembled through an interpolation scheme such as cubic splines or bilinear. Using the former parametric approach we can fit the surface to pass through all data points. However, the later approach is simpler and more efficient process used in Cartesian coordinates.

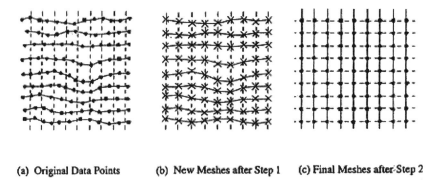

(a) Original Data Points   (b) New Meshes after Step 1   (c) Final Meshes after Step 2

Figure 5-8. Re-sampling process for mesh generation[6]

5.4.2 Digitization from the surface

A typical digitizing sequence for reverse engineering a body part (knee) is described here[14]. For a typical scan sequence, the CMM probe is traversed along a scanning line. Individual points are picked up through the scanning line. Figure 5-9 shows the scanning lines, which are the intersection of the scanning plane and the work piece surface. Both radial and axial scans are conducted. Before proceeding to the actual CMM scanning process, a measurement or inspection path planning is conducted. This inspection plan consists of pre-evaluating the positions of the scanning points, the direction of probing vectors, and a collision-free path for obtaining the best point cloud data. Choice of the right probe orientation is critical in picking up hard-to-reach points on the workpiece surface.

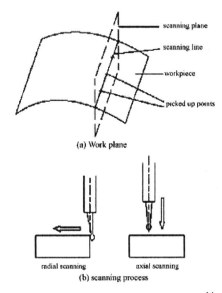

Figure 5-9. CMM scanning process[14]

5.4.3 Preprocessing of the point clouds

The quality of a 3D model is contingent on the preprocessing treatment of the point clouds. Procedures such as registration of point clouds in a single coordinate system and the removal of outliers ensure the completeness of the model[15]. The application of the CMM for measuring the artificial knee joint is explained in this section. The operator changes the measuring orientation in response to the feature of the artificial knee joint and measures the physical model at different positions such as 0°, ±45°, ±90°, ±135°, ±180°, and so on[14]. Limitations in the measuring probe and measurement method lead to outliers within the original point clouds. Their presence can lead to inaccuracies in the reconstructed 3D model. As a result, diagnostics tools should be applied to rectify these inaccuracies. An added advantage of contact-based systems is that they reduce the outlier point cloud data over non-contact based systems[14]. Figure 5-10 shows the elimination of outliers on the right middle side of the point cloud data after preprocessing.

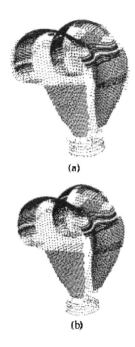

Figure 5-10. Point cloud before (a) and after (b) preprocessing[14]

5.4.3.1 Point processing as applied to a Knee Joint

The knee joint 3D model and its surface contour are composed of complex shapes rather than basic primitive features. In order to capture the point cloud of such a surface the knee is subdivided into different sections based on the basic primitives such as plane, cylinder, sphere, cone or a free form surface, etc. Each of these sections is called a block. Each block is scanned for point cloud data such that its interface can be clearly demarked or overlaps/intersects for further point processing. Each block of point cloud data is processed separately for higher precision of the resulting CAD model. Thus complex surfaces that cannot be directly scanned into a single mathematical surface can be captured using the subdivided primitive point cloud blocks[14].

5.4.4 Surface fitting

In order to create a 3D model the scanned-in point data from the physical model has to be converted from point to curve to surface to solid. This enables reproduction of individual surface feature from point cloud data. Proprietary software allow generation of high quality surface model with digitized point cloud data which lets the user compare and modify CAD models with point data, and speed up design iterations[14]. Several types of diagnosis methods can be applied to ensure precision in processing of points, curves and surfaces[14].

5.5 Performance Parameters of CMMs

5.5.1 Scanning Speed

The limitation in higher inspection speeds on CMM machines is related to the dynamic error[16]. The dynamic error value fluctuates at different regions of work zone based on the machine's inertia. The machine inertia is in turn dependent on the scanning speed and acceleration of CMM probe. Given the unpredictable and complex behavior of the dynamic error a clear cut methodology needs to be developed to minimize its adverse effects on the measurements. One of the simplest and easiest methods is to scan the work piece surface at relatively slow speeds such that the inertial effects are negligible. However, this strategy leads to longer scan cycles and higher scanning costs. A commercial CMM manufacturer (Renishaw) has developed an algorithm for scanning at faster speeds[16]. They rely on the fact that even though the dynamic error is unpredictable, it remains consistent across work pieces.

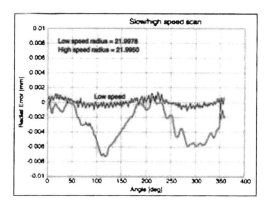

Here, the same circle is measured once at a slow speed and once at a fast speed. The differences between the two measurements show the CMM controller where the CMM's dynamic error lies.

Figure 5-11. Measurement accuracy during slow and fast speed scanning[16]

5.5.2 CMM Probe Accuracy[17]

One of most important factors influencing CMM accuracy is the type of probe used for measurements. A typical touch trigger probe outputs a binary signal when its tip is in contact with the object surface[17]. These signals are either generated by mechanical switching transducers or piezoelectric sensors. However, due to imperfections in the touch-trigger probe construction, the position of the probe during triggering is misaligned from the position of the real movement when in contact with the measuring surface. This misalignment is called "pre-travel". This pre-travel error can be determined using a master sphere of known radius. To determine the effect of the stylus ball and the average pre-travel, the most important parameter characterizing probe inaccuracy is the pre-travel variation. The most significant source of the touch trigger probe errors lies in the direction-dependent pre-travel variation.

Factors that influence the errors of touch trigger probes are categorized below[18]. The first type of factors consists of motion-related factors, such as a) speed of probe approach towards workpiece surface, b) probe acceleration, and c) approach distance. This group deals with the probe's impact with the workpiece surface.

The second group consists of probing design and configuration factors[17]. The probe errors depend on the a) the stylus mass and rigidity, b) the preload spring force, c) the probe orientation and d) the probe stylus length. The third group of factors is the method or mode of operation. Another group of factors is related to the operating environment. External events such as thermal drift, fluctuations in air temperature, ambient vibrations may be responsible for touch trigger probe errors. The last group of factors is directly related to the measured objects form, surface finish, and strength of probe material.

While operating the CMM under high probe speeds, it is possible that the probe transmits high acceleration values resulting in higher forces. These forces can create severely distort the probe stylus. In addition, if the probe is preloaded with a higher spring force, it can cause the stylus to deform during contact with the measured surface. All of the above reasons contribute to pre-travel variation of the probe.

5.5.3 CMM Rigid Body Errors[19]

A typical CMM has three perpendicular or orthogonal axes on which its components are mounted. Each axis consists of a linear stage which traverses along a single dimensional. However, due to finite stiffness and discrepancies in the manufacture and assembly of these stages, their mounted components undergo additional motion other than the theoretical one- dimensional motion. Considering rigid body kinematics of the mounted component, it has six degrees of freedom which include three translation motions (along X, Y and Z dimensions) and three rotational motions (around X, Y and Z axes). In addition, an out-of-plane motion exists for the components mounted on the linear stage. For a typical three-axis CMM, there are a total of 21 (3x6+3) parameters for the CMM probe location error within a rigid body kinematics model. The value of each of these parameters is dependent on the position of the component along the linear stage axis. One of the fundamental approaches to reducing rigid body errors is by high precision manufacture and assembly of the CMM structure and components. In recent years, an approach called "computer-aided error mapping" has been utilized for compensating these rigid body errors. The method consists of measuring individual rigid body errors at each location in the CMM workspace. This error is rectified either by moving the probe in real-time (active compensation) to the correct location or by mathematically evaluating the correct measurement coordinates once the measurement is done.

5.5.4 CMM Structural Deformations[19]

CMM distortions arise from a variety of sources including overconstrained mechanical systems, thermal gradients and rapid acceleration of carriage on the linear stage. As an example, the motion of an overconstrained carriage will result in the bending and twisting as it moves along the linear stage length. Similarly, rapid acceleration of the table or CMM probe towards the object to be scanned can cause temporary deformations within the linear guide structure. Errors due to the overconstrained carriage can rectified either by active or passive compensation techniques.

However, errors due to thermal gradients and rapid acceleration of components are transient in nature. One way to solve this issue is by installing active thermal sensors within the linear stage drives for continuous monitoring and compensation of thermal errors. A better method is to employ a temperature controlled environment for CMMs. Different parts of the CMM are made of components which have a wide range of thermal expansion coefficients. As a result there still exist local temperature gradients within temperature controlled environments. In the past, this error is addressed by the thermal error index method which combines a thermal drift test and an estimate of the uncertainty in the linear coefficient of thermal expansion[20]. However, this procedure may not give a complete picture of the structural distortions within the CMM. This calls for a detailed machine-specific experimental investigation of the effect of structural distortions on the accuracy of CMM measurements[21].

5.6 Integration of CMM data into other design and manufacturing system software

A typical CMM point-to-point or continuous surface scan generates point cloud data. This point cloud data set is further processed using curve fitting algorithms to generate a mesh of profile lines in the scanning directions. The next step involves the surface patch generation from the meshed lines. For accurate and rapid surface patch generation, two sensors are used to digitize an object[1]. Initially, a charged coupled device (CCD) camera is used as a low-level 3-D sensor to determine the spatial location of a part on the CMM bed and to recognize individual surface patches on the object's surface. A detailed scanning is done with the CMM touch probe to obtain precise 3D data. The simultaneous utilization of both the CCD camera and CMM touch probe offers benefits such as:

- Offline CMM touch probe tool path programming which saves valuable CMM machine time.
- Customized tool path can be generated for the touch probe based on the pre-selected surfaces of the workpiece.

Typical scanning procedure involves capturing of multiple stereo images for each side of the object. Different surface patches are integrated together to generate a solid model which can be converted into the universal CAD format such as the IGES. Both designers and manufacturers can use this CAD file for recreating the original part or modifying it for enhanced design. Rapid prototypes and simulation models can be built to test the proof of concept for a modified design. For creating a rapid prototype, the IGES file is converted into a triangulated file format called the Stereolithography (STL) format. Alternately, a tool path can be generated from the IGES file for CNC machining.

5.7 Recent Advances in CMM Technology

5.7.1 Reverse Engineering method based on Haptic Volume Removal[22]

Here, a new reverse engineering methodology based on haptic volume removal is described[22]. In this method, the object to be digitized is clad with virtual clay that is generated with a fixture. A haptic device with a position tracker is used to chip away the virtual clay. The popular commercial haptic device used in practice is called PHANToM®, from SensAble® Technologies[23]. A concurrent image of the chipped virtual clay is displayed on the computer monitor during the chipping operation. The operator gets a realistic feel of the chipping force from a force feedback haptic device. A haptic device is frequently used in conceptual design. A major benefit of this type of reverse engineering approach is that it does not need merging of point clouds generated from different views of current digitizers. The virtual clay volume is represented by a spatial run-length encoding scheme. The basic principle of a freehand haptic volume sculpting method is explained by a graphical illustration as shown in Figure 5-12[22].

Figure 5-12(a) shows a physical hand model to be digitized. Figure 5-12(b) shows two bounding boxes that are created by fixtures. These boxes encompass the hand model and simulate virtual clay. Figure 5-12(b) shows one box on the computer monitor. Another bounding box is invisible but can be sensed with the haptic device. The concept of delta volume is used for digitizing the physical hand. Delta volume is defined as the difference between the virtual clay and the physical object[22]. The object is digitized by chipping away the delta volume of the virtual clay with the position tracker of the haptic device.

(a) The physical hand

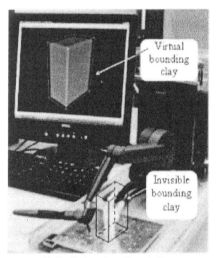

(b) Defining the virtual clay

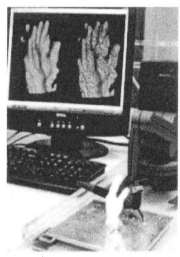

(c) Model emerging from virtual clay when sculpting

Figure 5-12. Haptic volume sculpting based reverse engineering[22]

Given the nature of physical contact and force application on physical models, this method is not suitable on soft objects. Another requirement is that the physical object should sustain chipping forces of the position tracker. Figure 5-12(c) shows the deformed delta volume, which is assigned a spring constant for haptic feedback. The force feedback is critical, because without it, the operator cannot feel the invisible virtual clay. The operator senses the un-sculpted virtual clay and proceeds towards a complete removal of the virtual clay. After complete removal of the virtual clay, a clear representation of the physical object is revealed. In Figure 5-12(c) the volume representation is shown in the left viewport on the screen and the surface model obtained with marching cube algorithms in the right viewport[22]. The haptic volume removal technique generates an unambiguous 3D volume model instead of the conventional point cloud data obtained from earlier digitization methods[24]. A six-degrees-of-freedom (DOF) PHANToM® haptic device with a sensing probe is attached for position and orientation sensing. Its design is analogous to the articulated type of digitizing systems.

The Haptic modeling technique is also used for a CMM inspection path planning environment. The Haptic Volume Coordinate Measuring Machine (HVCMM) simulates a real CMM in a virtual environment with haptic perception[25]. Operators can conduct offline programming tasks as if they are working with a real CMM. A collision-free inspection path can be generated by maneuvering the probe using a haptic device within a 3D computer-aided design (CAD) environment. The HVCMM generates a force feedback when the CMM touches the object to be measured. In addition, the operator can visually observe the contact phenomena, making it a faster and reliable off-line inspection method. The HVCMM not only speeds up the inspection path planning, but also can be used for the training of CMM operation.

5.7.2 Nano CMM[26]

Traditionally, CMMs have been used to measure coordinate locations of objects spanning measurement ranges from a few meters all the way to micrometer ranges. With the advent of Microelectromechanical Systems (MEMS) and Nanotechnology, it is essential that we have the ability to measure feature sizes in micrometer to nanometer ranges. Kiyoshi Takamasu, Professor, School of Engineering at the University of Tokyo, is developing a three-dimensional measuring instrument called a Nano CMM (Nano Coordinate Measuring Machine) for measuring objects in 3D space with nanometer resolution[26]. Microscopy techniques such as the Scanning Tunneling Microscope (STM) and Atomic Force Microscope (AFM) have been used for measuring the three-dimensional surface of objects at nanometer resolution. Though these microscopic techniques provide a wealth of topographical information, there are inadequacies, such as the length of a square notch formed in the side of an object, while measuring three-dimensional features. Figure 5-13 shows the construction of a Nano CMM[26].

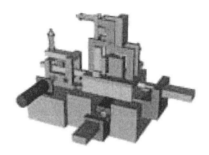

Figure 5-13. Construction of a Nano CMM[26]

Typical application for the three-dimensional Nano CMM is for measuring three-dimensional features and dimensions in the automobile industry. The fairly compact device has a resolution of 10nm and can measure feature sizes in the range of 10 x 10 x 10 mm. In order to achieve high stability, this measuring device has a symmetrical construction with a double V-groove guiding mechanism and employs a position sensing method using an optical scale. Figure 5-14 shows a prototype of the Nano CMM.

A significant source of measurement error at nanometer resolution ranges is the thermal drift. For example, when a 100-mm length of iron rises 1 degree in temperature, $1\mu m$ of thermal expansion occurs. This effect is directly reflected in the measured value[26]. The inventors claim that their prototype can restrain thermal drift of about 10 nm in an environment of about 0.1 degrees. This can be attributed to the usage of material with low thermal expansion coefficient which restrains any effects of temperature. A critical part of building a Nano CMM is the development of the probing system. Figure 5-15 shows the construction of a nanoprobe and a prototype.

Figure 5-14. Prototype of the Nano CMM[26]

The Nano CMM probe consists of a miniaturized ball with a diameter of less than 0.5 mm. In addition to the contact based measurement, an optical detector senses the movement of the ball. Thereby, when combining two techniques, high nanometer resolution measurements are attainable with the Nano CMM. This device will be an important tool for developing micro- and nano-machines. However, various issues such as reliability, durability, and calibration of this measuring instrument need to be evaluated before its field use.

Figure 5-15. Construction and prototype of a nanoprobe[26]

References

1. V. H. Chan., C. Bradley and G. W. Vickers, A Multi-Sensor Approach to Automating Co-ordinate Measuring Machine-based Reverse Engineering, *Computers in Industry*, **44**(1), 105-115 (2001).

2. S. Motavalli, Review of Reverse Engineering Approaches, *Computers ind. Engng*, **35**(1-2), 25-28 (1998).
3. J. A. Jalkio, Three Dimensional Inspection using Multistripe Structured Light, Optical Engineering, **24**(6), 966-974 (1985).
4. M. Idesawa, T. Yatagai, T. Soma, Scanning Moire Method and Automatic Measurement of 3-D shapes, Applied Optics, **16**(8), 2152-2162 (1977).
5. M. Chang, D. S. Wan, On-line Automated Phase –Measuring Profilometey, Optics and Lasers in Engineering, **15**(1), 127-139 (1991).
6. Ming Chang and Paul P. Lin, On-line Free Form Surface Measurement via a Fuzzy-Logic Controlled Scanning Probe, International Journal of Machine Tools & Manufacture, **39**(1), 537-552 (1999).
7. Renishaw, Inc., http://www.renishaw.com, Aug. 2, 2000
8. Renishaw, Inc., Instructional Presentation, University of Pittsburgh, Pittsburgh, PA, July 6-7, (2000).
9. Bidanda, Bopaya and Yasser A. Hosni, "Reverse Engineering and Its Relevance to Industrial Engineering: A Critical Overview," *Comput. Ind. Eng.*, **26**(2), 343-348(1994).
10. A. John. Bosch, Coordinate Measuring Machines and Systems, Marcel Dekker, Inc., (1995).
11. Bidanda, Bopaya, Vivek Narayanan, and Richard Billo, *Handbook of Design, Manufacturing, and Automation*, Ch. 48, "Reverse Engineering and Rapid Prototyping," John Wiley & Sons, Inc., New York, (1994).
12. http://www.coord3.com/web/en/cmm/cmm.html
13. http://www.lk-cmm.com/LKMSIWebSite/index.html
14. Yan-Ping Lina, Cheng-Tao Wanga and Ke-Rong Daib, Reverse engineering in CAD model reconstruction of customized artificial joint, Medical Engineering & Physics, **27**(1), 189–193 (2005).
15. Woo H, Kang E, Wang S, et al., A new segmentation method for point cloud data, International Journal of Machine Tools and Manufacturing, **42**(1), 167-217 (2002).
16. http://www.mmsonline.com/articles/0904rt1.html, Peter Zelenski, Faster Inspection On Existing CMMs, Rapid Traverse Technology and Trends Spotted By The Editors of Modern Machine Shop.[1]
17. Adam Wozniak and Marek Dobosz, Influence of measured objects parameters on CMM touch trigger probe accuracy of probing, Precision Engineering, (2005).
18. Johnson R. P, et al., Dynamic Error Characteristics of Touch Probes Fitted to Coordinate Measuring Machines, IEEE Trans Instrum Meas, **47**(1), 1168-1172 (1998).[1]

19. Steven D. Phillips, Chap 7. Performance Evaluation - Coordinate Measuring Machines and Systems, Contribution of the National Institute of Standards and Technology - NIST, Gaithersburg, Maryland, 159-165, (1995).
20. ANSI/ASME B89.6.2-1973 (R 1979), Temperature and Humidity Environment for Dimensional Measurement, ASME, Newyork, (1973).
21. P. S. Lingard, M. E. Purss, C. M. Sona, E. G. Thwaite, and G. H. Mariasson, Temperature Perturbation Effects in a High Precision CMM, Precision Engineering, **13**(1), 41-51 (1991).
22. Zhengya Yang and Younghua Chen, A Reverse Engineering Method based on Haptic Volume Removing, *Computer-Aided Design*, **37**(1), 45-54 (2005).
23. http://www.sensable.com
24. Yonghua Chen, Zhengyi Yang and Lili Lian, On the Development of a Haptic System for Rapid Product Development, *Computer-Aided Design,* **37**(1), 559-569 (2005).[1]
25. Y. H. Chen, Y. Z. Wang, and Z. Y. Yang, Towards A Haptic Virtual Coordinate Measuring Machine, *International Journal of Machine Tools & Manufacture*, **44**(1), 1009-1017 (2004).
26. K. Takamasu, Nano Coordinate Measuring Machine and Nanoprobe, Micromachine- MMC, No. 40 (2002).

Part II. Methods and Techniques

Chapter 6

VIRTUAL ASSEMBLY ANALYSIS ENHANCING RAPID PROTOTYPING IN COLLABORATIVE PRODUCT DEVELOPMENT

Kyoung-Yun Kim, Ph. D.[1] and Bart O. Nnaji, Ph.D.[2]
[1]*Department of Industrial and Manufacturing Engineering, Wayne State University, 4815 Fourth St., Detroit, MI 48201, kyoung_y_kim@yahoo.com,* [2]*US NSF Center for e-Design and Department of Industrial Engineering, University of Pittsburgh, Pittsburgh, PA 15261, nnaji@engr.pitt.edu*

Abstract:
This chapter discusses how assembly operation analysis can be embedded transparently and remotely into a service-oriented collaborative assembly design environment and how the integrated process can help a designer to quickly select robust assembly design and process for rapid manufacturing. A new assembly operation analysis framework, relevant architecture, and tools are introduced. True competitive advantage can only result from the ability to bring highly customized quality products to the market at lower cost and in less time. Product development has become a very complicated process. Many customers are no longer satisfied with mass-produced goods. They are demanding customization and rapid delivery of innovative products. Industries now realize that the best way to reduce life cycle costs is to evolve a more effective product development paradigm using the Internet and web based technologies.

Yet there remains a gap between these current market demands and current product development paradigms. The existing CAD systems require that a product developer possess all the design analysis tools in-house making it impractical to employ all the needed and newest tools. Instead of the current sequential process for verifying and validating an assembly design concept, a new Virtual Assembly Analysis (VAA) concept is introduced in this chapter to predict the various effects of joining to realize a rapid manufacturing environment. The predicted effects provide very important decision information to select a robust assembly design and to reduce unnecessary feedback processes on rapid selection of assembly processes.

Key words:
virtual assembly analysis; collaborative product development; computer-aided design; Internet; joining; assembly operation; welding; riveting; rapid prototyping; service-oriented architecture; *e*-tools; *e*-design; CAE; collaborative virtual prototyping and simulation; FEA; assembly design formalism; assembly analysis model; Pegasus service manager; *e*-design brokers; service providers

6.1. History and Overview

Generating over $1 trillion in annual revenue, the mechanical products industry is one of the leading industrial sectors in the United States economy. Throughout the sector, assembly operations account for a significant percentage of manufacturing activities. The U.S. Bureau of Labor Statistics indicates that those involved with assembly and welding processes exceed 2 million workers, which is more than 10 percent of the manufacturing workforce[1]. These numbers indicate that improving assembly operations will have an impact on manufacturing. While assembly processes in manufacturing industries are becoming mechanized and automated, the physical effects of assembly operations, particularly of welding, are still a major problem related to quality and productivity. The American Welding Society, in conjunction with the Department of Energy, has put together a vision that will carry the welding industry through 2020. They selected *assembly operation modeling, virtual simulation and testing*, and *proper joint determination* as key technology needs of industry.

Mechanical products are very rarely monolithic. Joints on a mechanical assembly are inevitable because of the limitations on component geometric configuration and material properties along with the requirements of inspection, accessibility, repair, and portability[2]. Very often joints are weak points from a mechanical or chemical viewpoint[3]. For example, many failures due to fatigue or corrosion occur at welded joints. From an efficiency viewpoint, joints often need some extra material to be added to the assembly structure, such as screws, bolts, or welding filler metal. The addition of extra materials sometimes leads to local weakening of the mechanical properties of the assembly components: for instance, the heat affected zone of a weld. To predict potential joint design problems, the trial and error procedure is generally used in assembly design process. The current design practice and analysis for verifying an assembly design concept is usually performed after selecting a final design concept. For example, a welding operation can generate thermal expansion and distortion of structure that will affect the joint and finally the entire structure. If a welded structure is distorted, then precision assembly cannot be achieved. Therefore, weld distortion should be minimized by optimizing the welding operation or by the use of an alternative joining method, such as joining with cast nodes. As another illustration, the rivet joints of an aircraft body frame should be capable of sustaining the prescribed load and/or mechanical forces in physically holding the assembly components together. If analyses indicates that stress level is not well balanced, the number of rivet joints could be optimized or an alternate joining method, such as welding, could be considered. The effects of joining operations, in the current design process, are analyzed after finishing assembly modeling. If analyses indicate that certain modification is required, then another iteration of modeling is needed. This process can be arduous and time-consuming.

The mechanical products industry requires a high level of performance in productivity and quality to maintain a competitive edge in the global economy. True competitive advantage can only result from the ability to bring highly customized quality products to the market at lower cost and in less time. Product development has become a very complicated process. Many customers are no longer satisfied with mass-produced goods. They are demanding customization and rapid delivery of innovative products. Industries now realize that the best way to reduce life cycle costs is to evolve a more effective product development paradigm using the Internet and web based technologies. Yet there remains a gap between these current market demands and current product development paradigms. For example, existing CAD systems require that a product developer possess all the design analysis tools in-house making it impractical to employ all the needed and newest tools.

This chapter addresses a methodology for assembly operation analysis to be embedded transparently and remotely into a service-oriented collaborative product development environment. This chapter also discusses how the integrated process of assembly can help a designer to quickly select a robust assembly design and process with the aid of virtual and physical prototypes. Instead of the current sequential process for verifying and validating an assembly design concept, a new Virtual Assembly Analysis (VAA) concept will predict the various effects of joining. The predicted effects provide very important decision-making information for selecting a robust assembly design and reducing unnecessary feedback processes during physical prototyping processes including rapid prototyping. The developed VAA framework provides a concurrent environment for designers to predict the physical effects transparently and remotely. The captured physical effects of assembly operations provide information critical to realizing an Internet-based collaborative product development environment. Finally in this chapter, the VAA framework, relevant architecture, and tools are implemented and applied to realistic assembly structures.

6.2. Modern Design and Product Development

6.2.1 Rapid product development

Rapid technological changes and increased globalization of markets continues to result in product proliferation as a major challenge to global companies[4]. Trends in vehicle design and assembly such as shared platforms and modules help reduce the complexity of final assembly[5]. The modular approach also can reduce the variation or the amount of inventory that is subject to uncertainty reduces, saving money[6]. An example of modular approach is delayed product differentiation (or commitment) introduced by Lee[7] in 1990s. Many researchers[4,8] showed the fact that delayed product differentiation can be viewed as a strategy for a company to improve the service level and reduce inventories when dealing with product proliferation.

Rapid prototyping is a technology for quickly creating physical models and functional prototypes directly from CAD models. Since the first rapid prototyping systems (stereolithography apparatus) emerged in 1988, rapid prototyping technology has been developing very fast[9,10]. The role of rapid prototyping has been highlighted as supporting product development processes (Figure 6-1)[9,11].

However, to rapidly develop a product and deliver its primary functions, the interactions between modules and assembly components should be well defined. The reliability of a product highly depends on its component interfaces. Different joining methods imply different physical implications. Often these different implications are not considered in rapid prototyping. These implications should be predicted proactively to prevent potential problems in modules and their interfaces in modern product design environment including rapid prototyping processes.

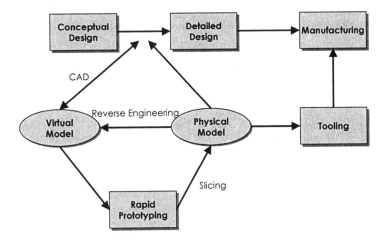

Figure 6-1. Rapid prototyping in product development (Adapted from Kochan *et al.* [11])

6.2.2 Internet-enabled collaboration

The impact of the Internet technologies has accelerated the pace of product development and product aftermarket service. Market competition has made quality not only an objective but also a prerequisite for companies to compete in the global marketplace. Smart companies must focus on product development and service innovation as a way to satisfy customers' need and lessen the cost and time associated with satisfying the customer. A new paradigm focusing on Internet–enabled tools (*e*-tools) for product design, rapid prototyping, manufacturing, and service is becoming a benchmark strategy for manufacturing companies to compete in the twenty-first century[12–17].

Many industries involved in the production of mechanical products including the aerospace, automotive, and medical devices industries, have taken the lead from the CAD system providers in developing strategic web-based benchmarking tools to globally integrate their product development. For example, General Electric Aircraft Engines in 2000 initiated a $21 million project funded by the US National Institute of Standards and Technology's Advanced Technology Program to develop web-enabled tools that ensure interoperability of design data among various CAD systems and among their OEMs[13]. There is now a realization by industries that product development tools need to be decentralized and be made accessible via the

Internet, and that technologies, which can allow for virtual prototyping and simulation, should be developed. While industries want to acquire these new technologies, they do not want to lose investment in existing design and analysis systems. There is a need for these systems to "talk" to one another, exchange needed data without transferring whole files, and become part of the new design structure and paradigm. Commercial CAD companies including PTC, SolidWorks, and IBM have shown strong interest in the integration of CAD and CAE environments[18-22]. These companies have developed their own integrated analysis tools. Nonetheless, their workbenches still require all tools to reside in-house and have a limitation on realizing a transparent and remote prototyping for collaborative product development.

There has been some research to realize rapid prototyping environments in distributed and Internet based environment. Fischer[23] proposed a multi-level CAD system to realize a remote reverse engineering environment. His systems utilized 3D faxing technologies. Luo et al.[24] presented Internet-based remote control and monitoring system for rapid prototyping. They combined rapid prototyping machines, preprocessors, and the Internet into a tele-control system. Luo and Tzou[25] developed an intelligent web-based rapid prototyping manufacturing system that combines a PC-based controller and the thermal-extrusion method using the web. Lin et al.[26] proposed scheduling approach for a web-based rapid prototyping manufacturing. They designed a distributed web platform where users can design prototypes, simulate manufacturing, and issue order-requests for real manufacturing through browser interfaces. Choi and Chan[27] developed virtual prototyping system that integrates virtual reality and rapid prototyping to facilitate virtual product development. They employed the dexel-based and the layer-based fabrication approaches to simulate the powder-based and the laminated sheet-based RP processes, respectively. Even though the previous research described above is meaningful efforts to realize rapid prototyping in the web or Internet, it still has limitations to differentiate different joining methods implying different physical implications in Internet-based collaborative product development. The virtual assembly analysis presented in this chapter enhances the rapid prototyping in collaborative product development by providing a transparent and remote assembly analysis capability.

6.2.3 Inevitable impact on assembly and joining operations

An assembly is a collection of manufactured parts, brought together by assembly operations to perform one or more of several primary functions.

An assembly operation is defined as the process or series of acts involved in actual realization of an assembly. Joining finalizes an assembly operation and generates joints. Messler[2] divided the primary functions into three categories: structural, mechanical, and electrical. Usually, assemblies perform multiple functions, with some function being primary and the others secondary. Therefore, the joints in an assembly also perform multiple functions. The primary function of the joints in an automobile frame is to provide a structural connectivity. They may also have a secondary function of allowing certain movement corresponding to vibration of the structure. To achieve function, diverse material properties and multiple parts are often employed. To enable material and structural optimization, determining appropriate joining methods and design is critical and can provide additional benefits, in terms of assemblablility, joinability, and quality and damage tolerance by changing properties along a potential crack path, and arresting crack propagation. Local joints should be compatible with the overall structure design. For example, if a deformation effect of a weld joint on a metal frame is propagated onto automobile windshield area, it can result in a fitting distortion problem between the window and the metal frame[28]. Understanding the physical effects of each assembly operation is very important for generating an appropriate joint design for an assembly.

As common joining methods, arc welding and riveting can be selected. In the arc welding process, due to the highly localized transient heat input from arc welding, considerable residual stresses and deformations, such as welding distortion, welding shrinkage, and welding warpage occur during heating and cooling in the welding cycle[29, 30]. Residual stresses are internal forces occurring without external forces. Plastic upsetting generated during heating is concomitant with strains. The stresses resulting from the strains incorporate and react to generate the internal forces. These forces cause the deformations (i.e., bending, buckling, and rotation). A common method of permanent or semi-permanent mechanical joining is riveting. Thousands of rivets may be used in the construction and assembly of many artifacts, such as airplanes, ships, and automobiles. Installing a rivet consists of placing a rivet in a hole and deforming the end of its shank by upsetting (heading). Sufficient compressive elastic energy must be stored in the components to ensure that the rivet is placed in tension by stress relaxation when the compressive forging pressure is released. The quality of the riveted joint depends on the preparation of the hole and the control of the punch pressure cycle. The rivet design should be determined by considering the required strength of the assembled joint, the required ductility of the rivet material, and the control of the forging process. There has been much research done to study analytically and numerically welding and riveting operations[31, 32]. However, these researchers have not presented a methodology to simulate the assembly operations transparently and concurrently

6.3. Collaborative Virtual Prototyping and Simulation

A virtual prototype is a computerized representation of assembly components that implies the testing and analysis of 3D solid models on computing platforms. Virtual prototypes should provide access to representations of all of the physical, visual, and functional characteristics of the actual physical device. They can be subjected to virtual testing to simulate full qualification tests prior to manufacture. This has the benefits of reducing management risk and potential engineering changes associated with a new manufacturing environment, provided that the model utilizes enough knowledge to make it a valid representation of the real world[33, 34].

Current Virtual Prototyping and Simulation (VP&S) has the following limitations: 1) complex and accurate physics-based virtual prototypes have traditionally required high-end and expensive workstations to host the software; 2) modeling and simulation software has been located on stand-alone computers with no opportunity for designers to perform collaboration activities; and 3) virtual testing is typically unavailable or must be performed through painstaking programming of simulation characteristics of a virtual prototype[35]. In addition, collaborators have limitations on how much information they can process at once.

To enable true collaborative VP&S, a virtual prototype should be generated with the consideration of various product life-cycle aspects and should be shared among distributed design collaborators[35]. To generate a robust virtual prototype, an understanding of assembly geometry and its physical effects is necessary. However, current solid modelers and simulation software are not really advantageous for driving a robust virtual prototype since they provide incomplete product definitions and are not able to act according to the semantics of the information. The reason for this is that traditionally the geometric model was in the center of product models. Of course, geometry is of primary importance in assembly design, but the morphological characteristics are consequences of the principle physical processes (e.g., deformation effects of welding) and the design intentions (e.g., joint intent)[36].

6.4. Service-oriented Collaborative Virtual Prototyping and Simulation

Collaborative VP&S can be realized in a service-oriented collaboration environment. Instead of looking at various engineering tools from traditional computation viewpoint, service-orineted collaboration focuses on the engineering implication of engineering tools from a more abstract level. This approach assures good openness of collaborative engineering systems. The

Internet is no longer a simple network of computers, rather from an application perspective, the Internet is a network of potential services. For example, in a three-tier web-based database system consisting of the web browser, server, and database, the web browser provides web document presentation services for human users, the web server provides data processing and retrieval services for the web browser, and the database provides data storage services for the web server. Thus, the Internet can be regarded as a complex system of service chains. Computer-aided design and engineering tools can be hooked up to the design platform over the Internet and provide certain services resulting in a distributed product development environment. This incorporates different engineering services and makes them available for automatic transactions in a collaborative product design environment. This product development environment is called an "*e*-design and realization environment."

The Internet provides an opportunity for these engineering tools to work together and utilize these services optimally. To connect these "islands of automation" transparently, universally accepted protocols should be defined at different levels. In a service-oriented collaboration, services should be specified from the functional aspect of service providers. To make an existing tool available online or to build a new tool for such a system, services associated with this tool should be defined explicitly. The service transaction among service providers, service consumers, and the service manager within the *e*-design system is illustrated in Figure 6-2. Once a service is registered at a central administrative manager, it is then available within the legitimate domain. This process is *service publication*. When a service consumer within the system needs a service, it will request a lookup service from the service manager. This process is *service lookup*. If the service is available, the service consumer can request the service from the service provider by the aid of the service manager. Most importantly, this service triangular relationship should be built at run-time. The service consumer (client) does not know the name, the location, or even the way to invoke the service from the service provider (server)[15, 16]. Collaboration between engineering tools is established and executed based on the characteristics of services that can be provided. In this case, different VP&S services may include type of analysis, visualization, computational resource, etc.

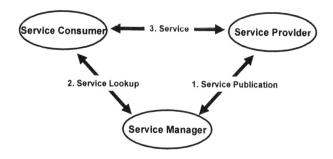

Figure 6-2. Service triangular relationship

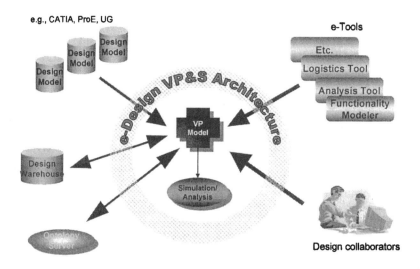

Figure 6-3. Collaborative virtual prototype model generation

Service-oriented architecture provides a fundamental infrastructure for collaborative VP&S. For collaborative VP&S, the virtual prototype should be generated with consideration of various product life-cycle aspects (e.g., functionality, manufacturing, assembly/joining, safety, ergonomics, material, packaging and shipping, and disassembly/recycling) and should be shared among distributed/remote design collaborators. Current virtual prototype models cannot fully capture multi-disciplinary data/information from collaborators or engineering tools in a distributed and heterogeneous collaboration environment. Also, to represent quality of a product, the virtual prototype should be capable of capturing realistic manufacturing situations. Figure 6-3 shows the concept of collaborative VP&S and virtual model generation based on service-oriented architecture. As shown in the figure, the information about different aspects of a virtual prototype model can be linked so that *e*-design collaborators/*e*-tools of various disciplines can access and manipulate it. This collaborative virtual prototyping architecture provides methods for sharing data/information between different collaborators/tools, allowing concurrent access to a virtual prototype model, and moving objects through the Internet.

6.5. Virtual Assembly Analysis

The current design and analysis practices for verifying a design concept are usually performed after selecting a final design concept. Prediction of various effects corresponding to specified assembly processes in up-front design is critical to understanding the performance of an assembly. VAA is a transparent and remote virtual simulation and testing paradigm utilized in a service-oriented collaboration environment. A VAA tool embedded in the assembly design process can be used to represent an assembly and imply the physical effects of a joint. Figure 6-4 illustrates the concept of VAA. A designer, who participates in service-oriented collaborative design, can request analysis services through the Internet/intranet. An analysis service provider solves the analysis problem requested and provides the results to the designer. This VAA process is embedded in the distributed assembly design environment and can guide designers to make appropriate design decisions. It generates an assembly design for joining and eliminates the time-consuming feedback processes between the assembly design and analysis processes.

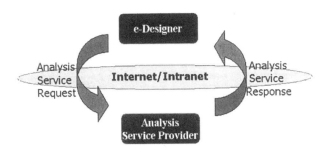

Figure 6-4. Virtual assembly analysis

6.5.1 Service-oriented VAA Architecture and Components

To realize VAA in a service-oriented architecture, an appropriate VAA service triangular relationship should be developed. In this triangular service relationship, each analysis service provider has its own service defined and published at the service manager. For example, an assembly design engine and VAA tool provide the services of assembly functional specification, engineering relations construction, and design presentation to end users. Many third-party analysis solvers can serve as the analysis service provider; the ANSYS solver provides the services of structural nonlinearities, heat transfer, dynamics, electromagnetic analyses, etc. During the process of service, one service provider may require services from other service providers. The first service provider will then send a service request to the providers, which provide these additional services. This service chain action should be transparent to the end consumer. For instance, when a design engineer completes the design of two parts, he/she may want to build an assembly model based on the part models. The detailed modeler then calls the assembly procedure. When the assembly model is finished, the design engineer may want to do further assembly analysis of the assembled parts by calling the service of a FEA tool through the VAA tool before expensive physical prototyping. The locations of various service providers are not known until run-time and the relation between the service consumers and the service providers is built dynamically. This relation can be viewed as a dynamic service chain, which connects service providers with client/server affiliation. Figure 6-5 illustrates the service-oriented VAA architecture. A designer, who participates in Internet-based collaborative assembly design, can request analysis services through the Internet/intranet. An analysis service provider solves the analysis problem requested and provides the results to the designer. As shown in Figure 6-5, the VAA architecture consists of four major service components: the VAA tool, Pegasus service manager, *e*-design service brokers, and service providers.

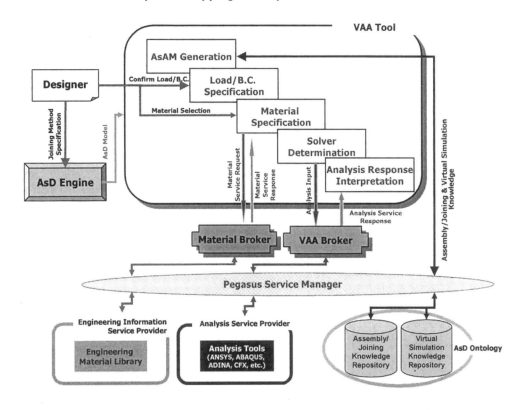

Figure 6-5. Service-oriented VAA architecture

6.5.2 VAA Tool

The VAA tool is an interface to VAA processes. When the designer wants to know physical effects of the specified joining, the VAA tool is triggered. If the designer doesn't possess any analysis tools in house, and/or has not any expertise in mechanical analysis, the designer can request services remotely and transparently by using this tool. The assembly operation analysis setup process is cumbersome and requires certain level of expertise. This process can be automated by imposing assembly/joining information on an assembly design model and extracting assembly analysis information from the assembly design model. Kim et al.[37] developed an assembly design formalism to persistently capture assembly/joining information in collaborative assembly design. The following sub-section briefly describes their assembly design formalism.

6.5.2.1 Assembly design formalism and assembly design model generation

The assembly design formalism specifies assembly/joining relations symbolically and it is used as the mechanism to perform product assembly design tasks. This assembly formalism is comprised of five phases: spatial relationship specification, mating feature extraction, joint feature formation and extraction, assembly feature formation, and assembly engineering relation construction. By interactively assigning *spatial relationships*, the designer can assemble components together to make final products and infer the degrees of freedom remaining on the components. In assigning *spatial relationships*, the mating features are defined and extracted from the parts. Mating feature extraction is a preliminary step to capturing joining information. This process provides geometric information directly related to assembly operation. However, the mating feature is not sufficient to represent a joining operation. The joint feature captures the information of actual joining operations. The designer can specify specific joining methods and constraints, such as welding conditions and fixture locations in joint features. After joint features are generated, assembly features are formatted. The purpose of assembly feature formation is to group the mating features and joint features together and thus integrate the data embedded at the component design stage with new assembly information for subsequent processes such as assembly violation detection, process planning, etc. Having designated spatial relationships, mating features, and joint features, the system can then trace back to the component design stage to determine from which design features these mating features originate and what their design specifications are. From the generated assembly features, assembly engineering relations, including assembly/joining relations, are automatically extracted and mating bonds are generated. A mating bond is a data structure representing mating pair and its mating conditions. Assembly engineering relations of the entire assembly are constructed based on the assembly features after specifying the spatial relationships and joining relationships between components. The mating bonds and an Assembly Relation Model (ARM) are used to represent the engineering relationships on the entire structure. A detailed description can be found in Kim *et al.*[37].

Kim *et al.* [38, 39] extended the assembly design formalism to represent assembly design (AsD) and joining information and constraints explicitly using ontologies. Their approach was to represent all of assembly design concepts explicitly and also in a universally acceptable manner. By relating concepts through ontologies rather than just defining data syntax (e.g., via XML), assembly/joining concepts can be captured in their entirety and extended as necessary. Furthermore, the higher semantic richness of ontologies allows computers to infer additional assembly/joining knowledge and make that knowledge available to AsD decision makers. By using ontology technology, AsD constraints can be represented in a standard manner regardless of geometry file formats. Such representation will significantly improve integrated and collaborative assembly development processes, including VP&S. Lastly, given that knowledge is captured in a standard way through the use of an ontology, it also can be retrieved, shared, and reused during collaboration. The developed AsD ontology forms an assembly knowledge repository.

From the AsD model, the VAA tool automatically generates an Assembly Analysis Model (AsAM) including the analysis variables, such as environmental variables, loading/boundary conditions, and material properties.

6.5.3 Assembly analysis model (AsAM) generation

To integrate assembly design and assembly operation analysis, assembly design models should be translated to an assembly analysis models. There have been some research investigations conducted to integrate product design and analysis. Peak et al.[40] presented a multi-representation architecture of intension of CAD-CAE integration. As an information-intensive mapping between design models to analysis models, a product model-based analysis model is researched and a framework to achieve design-analysis associativity is proposed. Rémondini et al.[41] developed a mechanical analysis module to generate an analysis data model from a geometric data model and mechanical information. Even though their methodology can be solutions for limited sense of CAD-CAE integration, they have not presented methods to integrate assembly design and assembly operation analysis to capture the physical effects of joining in the distributed assembly design environment. To perform VAA, the assembly/joining information necessary to assembly operation analysis can be extracted using an assembly-analysis solution model to explain physical phenomena based upon the assembly/joining information. The assembly-analysis solution model is an implantation of mapping functions (Ω) of assembly design and joining analysis. It translates an assembly design model (AsDM) to an AsAM: $_{AsDM}\Omega_{AsAM}$. Information essential for an assembly analysis is extracted from the AsD model, which is generated by the assembly design formalism.

The material of the assembly components from the joint features is translated to the material property for AsAM. Through this mapping, the material property for the specified material name is automatically assigned from a material library. If the resident material library doesn't have the information about the specified material, the material service can be invoked through the service-based architecture; a designer doesn't need to hold all material information in-house.

Heat input on a weldline, which is essential information to perform a weld analysis, can be calculated based upon assembly operation information, such as welding conditions (e.g., amperage, voltage, and welding speed) and material properties. Deposition of weld metal is simulated by defining the weld elements at elevated weld deposition temperatures. All other nodes are defined at the ambient temperature as the initial temperature field. Tables 6-1 and 6-2 show AsAMs for welding and riveting analyses. Information essential for an assembly analysis is extracted from the AsD model, which is generated by the AsD formalism.

Table 6-1. AsAM for welding

Assembly Design Model		Assembly Analysis Model
Information	Source Features	
Assembly Component Geometry	Assembly Feature	Geometry
Weld Geometry	Joint Feature	Geometry
Material Name	Assembly Feature	Material property
Fixture Location	Joint Feature (Joining Constraint)	Fixity location
Joining Conditions: welding condition, such as amperage, voltage, welding speed	Joint Feature (Joining Constraint)	Loading condition e.g., heat input

Table 6-2. AsAM for riveting

Assembly Design Model		Assembly Analysis Model
Information	Source Features	
Assembly Component Geometry	Assembly Feature	Geometry
Rivet Geometry	Joint Feature	Geometry
Material Name	Assembly Feature	Material property
Fixture Location	Joint Feature (Joining Constraint)	Fixity location
Joining Conditions: riveting condition, such as upsetting pressure	Joint Feature (Joining Constraint)	Loading condition e.g., pretension load

According to the assembly model information and assembly engineering information, additional geometric features, such as a weld bead for welded joints and a rivet for riveted joints, can be generated for detailed joint modeling. This detailed joint modeling provides realistic representation for engineering analyses. The configuration of the joint geometric features can be determined automatically from the assembly model information and assembly engineering information. For rivet joints, the designer specifies the location, head type, and radius of the rivet and the AsD model contains the information. For welded joints, the cross-sectional area of the weld bead can be determined from the existing theoretical relationships between the welding condition and the material properties imposed in the AsD model.

6.5.4 Pegasus Service Manager

The Pegasus service manager[15, 16] collaborates with third-party analysis servers (service providers), such as ANSYS, to achieve the VAA process. In this work, Common Object Request Broker Architecture (CORBA) is used to realize the service-oriented architecture for VAA. CORBA[42] is an architecture and specification for creating, distributing, and managing distributed program objects in a network. It allows programs at different locations and developed by different vendors to communicate in a network through an "interface broker". An ORB (Object Request Broker) acts as a "broker" between a client request for a service from a distributed object or component and the completion of that request. The ORB allows a client to request services from a server program or object without having to understand where the server is in a distributed network or what the interface to the server program looks like.

Service publication and lookup are the primary services provided by the service manager of VAA. Service publication includes *name publication*, *catalog publication* and *implementation publication*, which are provided for service providers. Name publication service is similar to the "white-page" service provided by telephone companies, by which the name of the service provider is published. Catalog publication service is similar to the "yellow-page" service where both the name and the functional description of the service provider are published. Implementation publication service is the procedure by which the service provider makes its implementation and invocation of services public so that clients can invoke the service dynamically. Correspondingly, service lookup includes *name lookup*, *catalog lookup* and *interface lookup*, which are for service consumers. Name lookup service is provided so that consumers can locate the service providers based on the service names. Catalog lookup service is for those consumers who need certain services according to their needs and specifications but do not know the names of the services. Interface lookup service provides a way such that consumers can check the protocols of how to invoke the service in the case that clients do not have the knowledge of the service in advance[15, 16].

Within the system, data transfers and transactions among servers can be completed based upon various distributed computing protocols, such as Hypertext Transfer Protocol (HTTP), Common Object Request Broker Architecture (CORBA), Distributed Component Object Model (DCOM), etc. Currently, the VAA architecture is implemented by CORBA. CORBA serves as a bond to integrate the whole system and provides good features of openness for collaborative computation. The components in the distributed system have peer-to-peer relationships with each other. From the end users' outlook, distributing application components between clients and servers does not change the look and feel of any single application, meaning, the system provides end users with a single system image.

6.5.5 *e*-Design Brokers

The *e*-design brokers handle service invocation and service result conveyance through Pegasus service architecture. The brokers reside in local sites; each client, such as the VAA tool, and each service provider needs the brokers to request or register service. The VAA tool can request the services by invoking these service brokers with relevant service inputs, such as analysis input files and material names. It minimizes the code modification of a service requesting system and provides plug-and-play capability. Before the VAA process, the analysis service providers register their service through an *e*-design broker at each server site. When VAA service is requested by a designer, the VAA tool sends a request with an analysis input to the *e*-design broker at the client site and the *e*-design broker conveys the request to the Pegasus service manager. After an analysis result is obtained from the analysis service provider, the Pegasus service manager informs the client's *e*-design broker and conveys the result to the designer.

6.5.6 Service Providers

The Pegasus service manager and the service providers play key roles in the VAA service chain management. The Pegasus service manager allocates service resources according to service consumers' demand and service providers' capability and capacity, while service providers respond the requested service.

In this chapter, two types of service providers are introduced: material service providers and analysis service providers. A specialized material service provider can provide the material properties, which are usually too cumbersome to store in the assembly designer's site. The designer can request a certain material properties from the engineering material service provider by specifying material name or certain material specifications. Any available engineering material library can provide relevant material properties to the client. To perform VAA to predict the physical effects of the joining, FEA tools, such as ANSYS, ADINA, and ABAQUS, can provide various FEA services. Generally, FEA tools allow certain command-based external analysis inputs. Depending upon the FEA tools and analysis types, different sets of commands and analysis procedures are needed. Appropriate analysis procedures, including specific analysis commands, can be provided from available analysis service providers through an analysis procedure service. In this study, typical analysis procedures considering the characteristics of joining methods are investigated and appropriate analysis procedures are pre-determined. Analysis service providers provide analysis procedure templates based upon the analysis procedures. The researchers at the NSF Center for *e*-Design are developing the assembly analysis knowledge repository based on ontology technology and this repository will be integrated with the AsAM.

6.6. Implementations

The VAA architecture and components are developed to realize the VAA process. This VAA process predicts the physical effects of joining processes, in which the VAA tool is embedded into assembly design processes and enhances rapid prototyping in collaborative product design environments. To realistically predict physical effect of joining, appropriate analysis procedures are required. This section describes examples of VAA procedures.

VAA for assembly operations requires specific analysis methodology and procedures. As a case study a thermo-structural analysis is used to understand the thermal and structural behavior of the welding operation. In addition, structural analysis is employed to predict various structural phenomena of the riveting operation. To enable VAA for specific joining processes, proper analysis procedures must be pre-investigated and built into an analysis procedure library.

To enable VAA, four major service components serve in the developed architecture: the VAA tool, transaction manager, service brokers, and third-party analysis service providers. In this case study, the Pegasus service manager is used as a transaction manager.

The VAA tool is implemented on ANSYS Workbench environment of ANSYS, Inc. The ANSYS solver is employed as the analysis service provider. Engineering material information is represented in XML format in the material database. The Pegasus service manager is implemented in Java. *e*-Design brokers are implemented in C++. IONA's ORBacus implementation of CORBA is used in the service architecture.

Figure 6-6 illustrates the transaction flow of services for VAA. Detailed processes are described below. The numbers in the figure stands for the index of each process.

STEP 1: Designers can exchange product data, such as AsD models, and select assembly components through the Product Data Sharing (PDS) service (①).

STEP 2: The selected assembly components are loaded in an AsD engine to generate joints (②). The system integrator, designer 1, can specify joining methods on the assembly (③).

STEP 3: When the designer wants to know physical effects of the specified joining, the VAA tool is triggered and a newly generated AsD model is sent to the VAA tool (④). From the AsD model, the VAA tool extracts analysis information and generates an AsAM. The designer can add additional loading and boundary conditions (⑤).

STEP 4: If the material specified in the AsD model doesn't exist in local database, the material property is obtained from remote material libraries though the service-oriented architecture. The designer can also request a certain material to be entered in the VAA tool. The VAA tool dynamically requests the service by invoking the material service broker (Mtl BK) with relevant material information (⑥).

STEP 5: Once the VAA inputs are ready, the VAA tool invokes the VAA service broker (VAA BK) with the VAA input. When the analysis is completed, the analysis service provider returns the analysis results to the VAA tool (⑦).

STEP 6: With the VAA results, the potential problem of AsD is fixed and the enhanced AsD is converted for rapid prototyping.

Rapid Prototyping: Theory and Practice 153

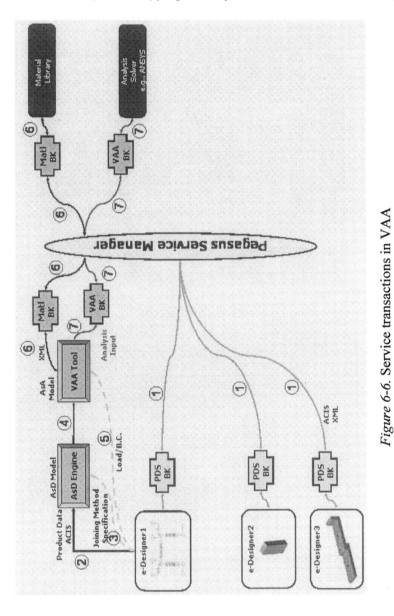

Figure 6-6. Service transactions in VAA

As shown in Figure 6-6, PDS service, material service, and VAA service are accomplished through service brokers (i.e., PDS broker, material broker, and VAA broker). These service brokers at the user's site handle service invocation and service result conveyance through the service-oriented architecture (Figure 6-6). The VAA tool can request the services by invoking these service brokers with relevant service inputs, such as analysis input files and material names.

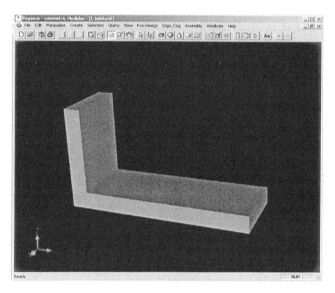

a) Assembly design model

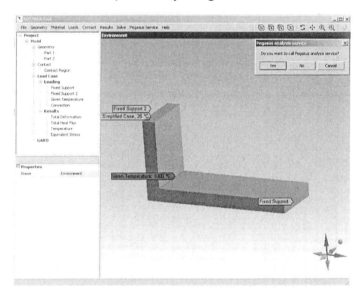

b) Assembly analysis model

Figure 6-7. Assembly models for VAA

The developed VAA tool (see Figure 6-7-a) is used as an interface to capture assembly and joining specification. When the designer wants to know physical effects of the joining, the VAA tool is triggered to interpret the AsD model. From the AsD models, the VAA tool automatically generates an AsAM including the analysis variables, such as environmental variables (e.g., as convection and fixed support), loading condition (e.g., given temperature and force/pressure), and material properties (e.g., Young's modulus, specific heat, and thermal expansion coefficient) (see Figure 6-7-b). The joining parameters (such as welding conditions and welding speed) are extracted from the AsD model and relevant analysis variables are obtained and assigned to the AsAM. For example, the degrees of freedom at fixture locations are restricted as fixed supports. Temperature at the specified weld seam is estimated from the welding condition. Through this analysis setup process, the designer can impose additional analysis constraints on the assembly analysis model in the VAA tool.

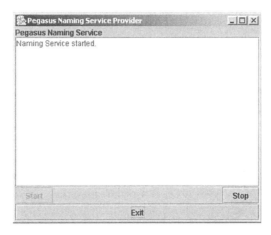

Figure 6-8. Pegasus Service Manager (Nnaji *et al.* [15, 16])

Figure 6-9. VAA service provider

The locations of various service providers are not known until run-time. The relation between the service consumers (such as the VAA tool) and the service providers is built dynamically. The Pegasus service manager allocates service resources according to service consumers' demand and service providers' capability and capacity. Figure 6-8 shows an implementation of the Pegasus service manager.

A specialized material service provider can provide the material properties, which are usually too cumbersome to store in the assembly designer's site. Here, an engineering material service provider has this information and offers engineering material lookup services. To perform VAA to predict the physical effects of the joining, the VAA tool (transparent to the analysis service provider) looks up and acquires the material information on the specified material type from the remote engineering material service provider.

Once complete the AsAM is generated, and VAA service can be invoked. VAA input for available VAA service providers is generated by the VAA tool considering specified joining method's characteristics and analysis preferences. For example, if the designer wants to perform a thermal analysis for the welded joint, the tool can generate appropriate inputs for the available VAA service provider to perform the thermal analysis. For this example, predetermined analysis procedures are used for VAA. Determining appropriate analysis procedures is very important for obtaining realistic analysis results. The service-oriented architecture provides an environment in which new analysis procedures are easily acquired from remote analysis service providers. This VAA service, which is provided by the analysis service providers (Figure 6-9), such as ANSYS, ADINA, and ABAQUS, can be invoked remotely through the Pegasus service manager. When the analysis is completed, the analysis service provider (see Figure 6-9) returns the analysis results (e.g., output files, animation movies) to the VAA Broker, eventually to the VAA tool (Figure 6-10).

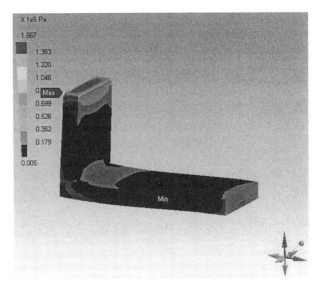

Figure 6-10. Equivalent stress and deformation obtained from VAA service

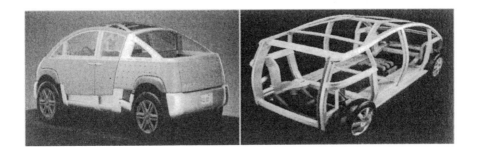

Figure 6-11. Aluminum concept car and body frames (Buchholz [43])

The VAA framework is also used for realistic examples, such as an aluminum space frame assembly of an automobile (Figure 6-11) and a hinge assembly with rivet joints (Figure 6-13). The welded frame[44] is made up of thin walled aluminum beams with rectangular sections and flat planer sections. Aluminum alloy (such as 6061 or 6063) extrusions have been considered as materials. Recent emphasis on lightweight environmentally sound car design has opened up the possibility of substituting lower-density corrosion-resistant recyclable aluminum for steel in car bodies[44]. However, the high distortion of aluminum alloy is a difficult problem to overcome to achieve precision manufacturing. For example, aluminum alloys 6061-T6 and 6063-T6 have a deformation index of 0.01 (worse) against an index of 1.0 for mild steel[45].

158   Rapid Prototyping: Theory and Practice

Due to the size of the car frame structure, the generation of a rapid prototype is expensive process. Also, the rapid prototype often cannot provide physical effects of joining (in this case arc welding). Figure 6-12 illustrates a simple car frame model in the VAA tool and the VAA result. The result clearly shows deformation of this structure and stresses concentrated at the welded joint. Deformation beyond allowable tolerance can be indicated easily. Based on this result, the designer can make a decision on whether this joining method is feasible before expensive physical and rapid prototyping.

As another illustration, a hinge assembly with rivet joints is used. The material of the hinge is a structural steel. Figure 6-13 shows a structural VAA result for the hinge joint. Based upon these results, the designer can clearly see that this joint is robust in this specific test environment.

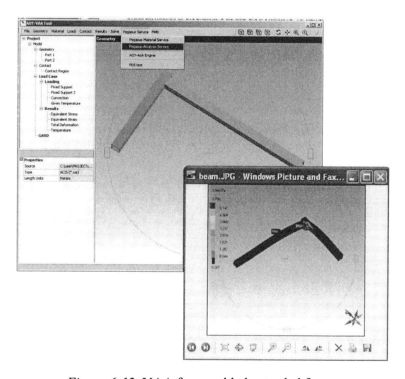

Figure 6-12. VAA for a welded extruded frame

Rapid Prototyping: Theory and Practice 159

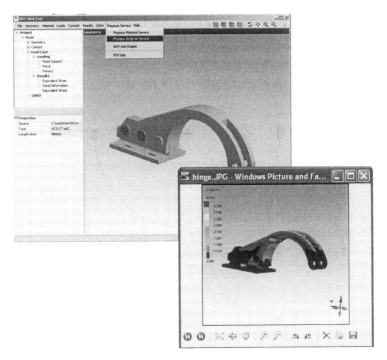

Figure 6-13. VAA for a hinge with three rivets

6.7. Contributions

The contribution of the VAA paradigm and tools are summarized below.

1. VAA can simulate products made with material that rapid prototyping cannot utilize economically. VAA provides a remote and transparent assembly analysis environment. Rapid prototyping can be utilized to test physical interfaces based on the VAA outcomes. When CAD models of other components (e.g., components contacting with the hinge assembly) are not available, rapid prototyping is more desired to be used.

2. VAA simulates physical effects of different joining process. Rapid prototyping is a material-dependent process. It is very expensive to generate prototypes made of final materials.

3. VAA allows for more iterations of testing design. When a new design feature (e.g., additional fasteners) is required and included, rapid prototyping cannot be performed right away and will require

entire new prototyping process. With VAA the designer can perform "what-if" type simulation.

6.8. Conclusions

In this chapter, an innovative assembly analysis framework, *virtual assembly analysis* (*VAA*), architecture, and components are introduced to predict the physical effects of selected joining processes and to enhance rapid prototyping in an integrated, collaborative product development.

To rapidly develop a product and deliver the primary functions of the product, the interactions between modules and assembly components should be well defined. The VAA tool and architecture can be utilized to differentiate joining methods implying different physical implications in modern design environment including rapid prototyping. Predicting these implications proactively can prevent potential problems in modules and their interfaces in product design.

References

1. AWS, Vision of Welding Industry, American Welding Society, http://www.aws.org/research/vision.pdf (2003).
2. R. W. Messler, Joining of Advanced Materials (Butterworth-Heinemann, 1993).
3. C. LeBacq, Y. Brechet, H. R. Shercliff, T. Jeggy, and L. Salvo, Selection of joining methods in mechanical design, Materials and Design, **23**, 405-416 (2002).
4. H. L. Lee, and C. S. Tang, Modeling the costs and benefits of delayed product differentiation, Management Science, **43** (1), 40 – 53 (1997).
5. 3daycar, http://www.3daycar.com/
6. D. E. Whitney, Mechanical Assemblies: Their Design, Manufacture, and Role in Product Development, Oxford Series on Advanced Manufacturing, Oxford University Press (2004).
7. H. L. Lee, Effective inventory and service management through product and process redesign, Operations Research, **44** (1), 151 – 159 (1996).
8. G. Barbiroli and A. Focacci, Product diversification in the vehicles industry: a techno-economic analysis, Technovation, **23**, 461 – 513 (2003).

9. D. T. Pham and S. S. Dimov, Rapid manufacturing: the technologies and applications of rapid prototyping and rapid tooling, Springer-Verlag (2000).
10. Y. Ding, H. Lan, J. Hong, and D. Wu, An integrated manufacturing system for rapid tooling based on rapid prototyping" Robotics and Computer-Integrated Manufacturing, **20**, 281 – 288 (2004).
11. D. Kochan, C. C. Kai, and D. Zhaohui, Rapid prototyping issues in the 21st century, Computers in Industry, **39**, 3–10, (1999).
12. J. Welch, Jack Welch and GE, Business Week, Oct. issue (1996).
13. FIPER, National Institute of Standard Technology Annual Review on FIPER, General Electric Aircraft Engines, Springdale, OH, December 12-13 (2001).
14. Pegasus, NSF IUCRC for *e*-Design: Strategic Planning Meeting, Pittsburgh, PA, USA, Dec 9 to 10, www.e-designcenter.info. (2003).
15. B. O. Nnaji, Y. Wang, and K. Y. Kim, Cost-Effective Product Realization – Service-Oriented Architecture for Integrated Product Life-Cycle Management, 7th IFAC Symposium on Cost Oriented Automation, Gatineau/Ottawa, Canada, June 7-9, plenary lecture (2004).
16. B. O. Nnaji, Y. Wang, and K. Y. Kim, Service-Oriented Architecture for Integrated e-Design and Realization of Engineered Products, International Forum on Design for Manufacture and Assembly, Providence, RI, June 22-23, keynote paper (2004).
17. H. Lan, Y. Ding, J. Hong, H. Huang, and B. Lu, A web-based manufacturing service system for rapid product development, Computers in Industry, **54**, 51–67 (2004).
18. OneSpace, CoCreate Corporate, http://www.cocreate.com
19. Windchill, Parametric Technology Corporate, http://www.ptc.com
20. Smarteam, Dassault Systems, http://www.3ds.com
21. Teamcenter, EDS, http://www.eds.com
22. CATIA, Tolerance Analysis for Flexible Assembly, CATIA, Version 5 Release 9, Training material.
23. A. Fischer, Multi-level models for reverse engineering and rapid prototyping in remote CAD systems, Computer-Aided Design, **32**, 27–38 (2000).
24. R. C. Luo, J. H. Tzou, and Y. C. Chang, An Internet-based remote control and monitoring rapid prototyping system, Proc. of the 27th Annual Conference of the IEEE Industrial Electronics Society (2001).
25. R. C. Luo and J. H. Tzou, The Development of an Intelligent Web-Based Rapid Prototyping Manufacturing System, *IEEE Transactions on Automation Science and Engineering*, **1** (1), 4 – 13 (2004).
26. H. H. Lin, C. W. Hsueh, and C. H. Chen, A real-time scheduling approach for a web-based rapid prototyping manufacturing platform,

Proc. of the 23rd International Conference on Distributed Computing Systems Workshops (2003).
27. S. H. Choi and A. M. M. Chan, A virtual prototyping system for rapid product development, Computer-Aided Design, **36**, 401–412 (2004).
28. B. O. Nnaji, D. Gupta, and K. Y. Kim, Welding distortion minimization for an aluminum alloy extruded beam structure using a 2D model, ASME Journal of Manufacturing Science and Engineering, **126** (1), 52 – 63 (2004).
29. K. Masubuchi, Analysis of Welded Structures: Residual Stresses, Distortion, and their Consequences, 60-236 (Pergamon Press Inc., 1980).
30. H. S. Moon and S. J. Na, Optimum design based on mathematical model and neural network to predict weld parameters for fillet joints, Jr. of Manufacturing Systems, **16** (1), 13-23 (1997).
31. C. L. Tsai, S. C. Park, and W. T. Cheng, Welding Distortion of a Thin-Plate Panel Structure, Welding Journal, 156-165 (1999).
32. Rahman, *et al.*, Boundary correction factors for elliptical surface cracks emanating from countersunk rivet holes, *AIAA Journal*, **38** (11), 2171-2175 (2000).
33. M. J. Pratt, Virtual prototypes and product models in mechanical engineering, Proc. IFIP WG 5.10 on Virtual Environments and their Applications and Virtual Prototyping, 113-128 (1994).
34. C. K. Chua, S. H. Teh, and R. K. L. Gay, Rapid prototyping versus virtual prototyping in product design and manufacturing, Int. Jr. of Advanced Manufacturing Technology, **15**, 597-603 (1999).
35. J. S. Lombardo, E. Mihalak, and S. R. Osborne, Collaborative Virtual Prototyping, Johns Hopkins APL Technical Digest, **17** (3) (1996).
36. K. Y. Kim, D. G. Manley, B. O. Nnaji, and M. R. Lovell, Framework and Technology for Virtual Assembly Design and Analysis, 2005 IIE Annual Conference (IERC), Atlanta GA, May 14 – 18 (2005).
37. K. Y. Kim, Y. Wang, O. S. Muogboh, and B. O. Nnaji, Design formalism for collaborative assembly design, Computer Aided Design, **36** (9), 849-871, (2004).
38. K. Y. Kim, D. G. Manley, and H. J. Yang, The Role of Ontology in Collaborative Virtual Prototyping, 2005 IIE Annual Conference (IERC), Atlanta GA, May 14 – 18 (2005).
39. K. Y. Kim, D. G. Manley, H. J. Yang, and B. O. Nnaji, Ontology-based Virtual Assembly Model for Collaborative Virtual Prototyping and Simulation, the 2005 International Symposium on Collaborative Technologies and Systems (CTS 2005), Saint Louis MO, May 15-19 (2005).

40. R. S. Peak, R. E. Fulton, I. Nishigaki, and N. Okamoto, Integrating Engineering Design and Analysis Using a Multi-representation Approach, Engineering with Computers, **14**, 93-114 (1998).
41. L. Rémondini, J. C. Léon, and P. Trompette, High-level Operations Dedicated to the Integration of Mechanical Analysis within a Design Process, Engineering with Computers, **14**, 81-92 (1998).
42. J. Siegel, CORBA 3: fundamentals and programming, 2nd ed. (John Wiley & Sons, Inc., 2000).
43. K. Buchholz, Alcoa shows aluminum association its concept vehicle, Automotive Engineering International, 53-55 (1999).
44. S. Ashley, Contour: the shape of cars to come?, Mechanical Engineering, **113** (5), 36-44 (1991).
45. D. Radaj, Heat Effects of Welding, 21-23 (Springer-Verlag, 1992)

Chapter 7

Subtractive Rapid Prototyping: Creating a Completely Automated Process for Rapid Machining

Matthew C. Frank
Department of Industrial and Manufacturing Systems Engineering
Iowa State University, Ames, IA 50011, mfrank@iastate.edu

Abstract:
　　This chapter presents a description of how CNC milling can be used as a rapid prototyping process. The methodology uses a layer–based approach for machining (like traditional rapid prototyping) for the rapid, automatic machining of common manufactured part geometries in a variety of materials. Parts are machined using a plurality of 2½-D toolpaths from orientations about a rotary axis. Process parameters such as the number of orientations, tool containment boundaries and tool geometry are derived from CAD slice data. In addition, automated fixturing is accomplished through the use of *sacrificial support* structures added to the CAD geometry. The chapter begins by describing the machining methodology, and then presents a number of critical issues that affect making the process automatic and efficient. The CNC-RP process is compared and contrasted to existing RP processes. In particular, we consider the differences in an additive versus subtractive process with respect to accuracy and material choices. The strengths and limitations of rapid machining are illustrated, along with a discussion on the economics of

using rapid machining versus additive RP and/or traditional machining processes to create single or small batches of parts.

Key words:

Rapid Prototyping, CAD, CAM, CNC Machining, STL, Process Planning

7.1. Background

Over the past two decades, numerous *additive* RP processes have been developed. The first commercially available process was Stereolithography, which uses photo-curable resin to create parts layer-by-layer from a vat of liquid. Since then, many processes that utilize layer-based principles have been developed, and today the price of low-end RP systems has decreased considerably. In particular, desktop 3D printing systems that can create concept models have become more commonplace. These systems create *Form* models, which simply represent the CAD geometry in physical form, although they are not useful for much more. Parts are created in weak materials with poor accuracy, however the processes are automated and quite simple to use. On the other end, much work has focused on developing systems that can create truly functional prototypes in strong materials. In particular, new systems were designed to work with metals, such as Direct Metal Laser Sintering (DMLS), Laser Engineering Net Shaping (LENS) and Electron Beam Melting (EBM). The motivation behind these systems is to create truly functional metal parts. In the previous RP systems, the commonly used materials are plastics, paper and powders, which is sufficient for concept models or perhaps plastic injection molded parts, for example. The biggest challenge with metals to date has been in the energy required to create layer based metal parts (e.g.: it is easy to sinter plastic particles, but more difficult to sinter metal). Since sintering cannot produce fully dense parts, systems like LENS use more powerful lasers and a stream of powder while EBM utilizes an electron gun to melt the powder layers. Obviously, the cost of these systems is significantly higher than others and can approach $1 million (US). Unfortunately, LENS and EBM often do not have the accuracy and surface finish capabilities required for the parts they are intended to create. Therefore, post-processing (typically using machining) is common, in order to provide better surface finishes and/or tight tolerances on critical features.

However, little serious attention has been given to exploring a push-button CNC machining process, or at least not within the context of an RP system. Although CNC machining has received tremendous attention in the literature, the focus has been on creating automated systems for *production* processes. However, we will argue that the motivation behind an RP system is often entirely different from a production system. In a production environment, it is the time spent producing each part that is important, while for an RP part the cost is in the time spent planning the process. In other words, if the production plan calls for 100,000 parts, then spending a week designing a set of fixtures, NC code and other process plans makes sense if the result is a very fast processing time for each part. However, if we only need to make one part, then we may be willing to allow the process to take a few hours. This would be acceptable if we only need to spend a few minutes planning, or moreover, if we just had to push a button. This has been proven true (at least empirically) by the success of current RP systems since they are essentially push-button processes, but are by no means "rapid" in their processing. Users do not care if the machines run for hours, overnight, or even a day or two since all they needed to do was upload an STL file and make a few clicks of the mouse.

So, why have we not taken the same approach with CNC machining? If we allow ourselves to extend the processing time for machining (as needed) can a creative set of approaches for process planning be developed? In this chapter, we describe a methodology that does just that, not to say that we try to increase the machining time, but we have only focused on the complete automation of the process planning steps. The goal of this work is to create a system for rapid CNC machining that is useable by *anyone*, not just skilled machinists. The objective is to create a set of algorithms and generic hardware that can take as input a simple CAD model such as an STL file, and then be able to directly send NC code to a milling machine to begin processing a part. The method described in the following chapter borrows from some of the general ideas behind other RP processes. For example, using a layer based (2½-D) approach to toolpath planning is found to be an easy way to create collision free toolpaths with no need for feature recognition. Also, we describe a method for using sacrificial supports to fixture the parts automatically. This is similar to RP systems like SLA and FDM, which add small structures under overhanging layers to support the part. However, in the rapid machining process, these structures are what are left holding the part when the machining process completes, rather than entities that are added to the part. One will see that this new method uses the best of both processes (RP and typical CNC machining) in a manner that allows the process to be completely automated, yet still provide the material flexibility and accuracy that is lacking in the additive RP processes.

7.2. Related Work

The typical approach to planning parts for CNC machining has been to define the "features" on the part, and match these features and tolerances to a set of processes that can create the required geometry to the specified accuracy. This approach has worked reasonably well for medium to high volume parts, but it has had marginal success for the production of very small quantities of parts. In most cases, the time required to plan the part, kit the required tooling, and setup the machine (both fixture and tooling) has limited the use of CNC for these applications. In the literature, process planning is often approached with a set of goals driven by high production levels of parts. That is, a set of plans that strive for cost-effectiveness through maximizing feeds and speeds and creating repeatable setups that can be paid for through economies of scale. Process planning for CNC machining includes tasks such as fixture planning, toolpath planning, and tool selection. There is a considerable amount of work in the literature pertaining to these three areas[1-3]. The concept of *flexible fixturing* has been the topic of much research, though a completely automatic fixture design system has yet to be developed[4].

Some exploration into the use of CNC machines for rapid prototyping has been published. Chen and Song describe layer-based robot machining for rapid prototyping using machined layers that are laminated during the process[5]. The process is demonstrated using laminated slabs of plastic, machined as individual layers upon gluing to the previous layers.

A hybrid approach using both deposition and machining called *Shape Deposition Manufacturing* (SDM) continues to be developed[6]. For each layer, both support and build material is deposited and machined in a combined additive and subtractive process. Sarma and Wright (1997) presented Reference Free Part Encapsulation (RFPE) as a new approach to using phase-change fixturing for machining[7]. The approach is discussed recently in conjunction with high speed machining (HisRP)[8]. RFPE, in combination with feature based CAD/CAM was proposed as an RP system[9]. Another approach is to use CNC machining for prototyping dies, an area called *Rapid Tooling*[10]. One approach to rapid tooling uses machined metal laminates stacked to form dies[11,12].

Many of these methods utilize CNC machining, but do not address the fundamental problems of automating a fully subtractive rapid machining approach. This chapter presents a method for "feature-free" CNC machining that requires little or no human-provided process engineering. The methodology described in this chapter is a purely *subtractive* process that can be applied to any material that can be machined. The method described herein was developed in response to the challenge of automating as much of the process engineering as possible. The ultimate goal is to generate both the NC code and an automatically executed fixturing system by the touch of a button, using only a CAD model and material data as input. The process is perfectly suited for prototypes as well as parts that are to be produced in small quantities (~1-10).

7.3. Assumptions

Before beginning a discussion on the methodology, it is necessary to elucidate the set of constraints, both known problems in CNC machining and some self-imposed by the researchers. For one, there will be a general assumption about the user; in particular, that the human planner has no experience in machining. This is justified in light of the fact that one use of this methodology is for prototyping. During the early stages of design, one cannot assume that an experienced machinist is available. The existing RP processes allow users to download CAD files and simply push a button to initiate the part building process. The same will need to be true for a rapid machining process. What does this mean in terms of the typical steps for process planning? The implication is that even moderately skilled tasks, including setting fixture and tool offsets, must be eliminated. More importantly, fixture design and implementation must be at least semi-automated. Overall, it is expected that the user will only be responsible for loading a piece of stock in a work holding device that is straightforward to use (e.g.: simple vice, chuck, collet, etc.).

Another assumption is that feature information will not be available as data input. In some cases of simple prismatic parts, feature extraction may not be a problem; however, for the general free form part shape, it cannot be assumed that accurate and complete feature information is known. An example would be a CAD model generated by laser scanning of biological objects, such as bones. This assumption suggests that toolpath plans must be generated without knowing what type of surface geometries are to be machined. This includes choosing the tool diameter and length, depth of cut, feeds and speeds. Specifically, this assumption implies that the process planning method is not intended for *populating* features on a piece of stock material, rather, the entire surface of the CAD model must be cut *from* the stock material. In other words, process planning does not have to be done for each feature individually and then each feature milled in a sequence of operations. Although the difference may appear subtle, this assumption will be shown to have a significant impact on the framework of this methodology and will be explored in further detail in the proceeding sections of this chapter. Lastly, it will be assumed that this process will be executed in a lights-out operation. Given that, any catastrophic failure such as crashing the tool, holder, or spindle with any part of the machine tool or fixture must be prevented.

7.4. Overview of the CNC-RP Process

This section presents a general overview of the current methodology. Methods have been developed to cover all aspects of process planning for rapid machining, including toolpath planning, choosing tool geometries, calculating setup orientations, and a concept for a universal approach to fixturing.

With regard to general toolpath planning, the CNC-RP method borrows from the layer-based RP technologies. The basic concept is to machine the visible surfaces of a part from each of a plurality of orientations. In order to simplify the problem from both a process and fixture-planning standpoint, only rotations about one axis for orientations of the stock material during processing are used. This not only reduces the problems associated with process planning, but it will be shown how this assures the absolute collision-free nature of the approach. From each orientation, some, but not all of the part surfaces will be visible. The goal is to machine the part from enough orientations, such that, after all toolpaths are complete, all surfaces have been fully machined from at least one orientation. For each orientation, there is not a particular plan for a set of *feature* machining operations; rather, it is machined using simple 2½-D layer-based toolpaths. Unlike the existing rapid prototyping methods, CNC machining is a *subtractive* process; therefore, one can only remove the material around the periphery of a part (visible cross section of the part). This is similar to *roughing* operations in traditional machining process planning, only in this case, the layer depth is set shallow enough (typically 20-120microns (0.001" to 0.005")) that one can expect *finish* machining results. Part surface contours are created with the same "staircase" effect seen in other RP methods. However, since machining is able to make very shallow depths of cut, rapid machining can produce very thin layer thicknesses. Rapid machining can achieve layer thicknesses easily down to 20microns (0.001") or less and can be varied depending on the requirements, not limited like other processes.

The *feature-free* nature of this method suggests that it is unnecessary to have any surface be completely machined in any particular orientation and this simple concept is illustrated in Figure 7-1. In the first operation (Figure 7-1a) much of this surface is visible from the first orientation; however, the dark areas under the overhanging surface are not visible. In the second operation, this originally "shadowed" region of the same surface is now visible (Figure 7-1b). This approach avoids the problem of feature recognition and feature-based process planning. At least two, but more likely numerous orientations will be required in order to machine all the surfaces of a part about one axis of rotation (even a simple part like a sphere requires two orientations).

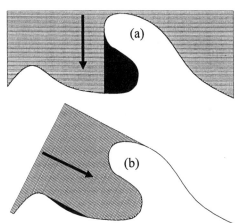

At each orientation, a tool is needed that can reach to the last layer for machining without colliding with any previously machined surface or the stock material. This requires that the shank diameter be less than or equal to the flute diameter. Since simple 2½-D toolpaths are being used, then a flat-end tool is an appropriate choice. Lastly, the diameter of the tool is simply the smallest diameter available in the given length. If it is assumed

Figure 7-1. Free-form surface being machined from two orientations

that no feature information is available, then the approach is to use a small diameter tool such that the most general shapes can be accessed. Unfortunately, there are tradeoffs with using the smallest diameter tool. For one, a small tool does not remove as much material as a tool of larger diameter for a given feed rate and depth of cut. The other problem is that small diameter long tools can be deflected more easily under cutting forces. These problems make it necessary to maintain feed rates and depths of cut such that the tool does not chatter, bend or break. Since we are using such small tools, the method may not appear very efficient; however, one will note that it is not actually important to find an efficient solution for a rapid prototyping process. Most RP processes are not very *efficient*, or at least not when the actual model building process is occurring (hours or days of stacking thin layers?); however, the process planning and setup MUST be fast, automatic and easy.

One would note that if all the visible surfaces of a part from numerous orientations were machined completely, then at some point the part would simply fall from the stock material. Therefore, this method employs a fixturing approach that is similar in concept to the *sacrificial supports* used in many existing additive rapid prototyping processes. However, in CNC RP the supports are not added to the physical model, rather, they are added to the CAD model prior to toolpath planning. The sacrificial supports are currently implemented as small diameter cylinders added to the solid model geometry parallel to the axis of rotation. During processing, the supports are created incrementally, along with the rest of the part surfaces. Upon completion, the finished part is left secured to the round stock material by these supports.

The rapid machining process is based on a setup strategy whereby a rotary device is used to rotate round stock material that is fixed between two opposing chucks. Rotating the stock using an indexer eliminates the inherent problem of retaining reference coordinates associated with re-clamping a part in a conventional fixture. For each orientation, all visible surfaces are machined and the sacrificial supports keep it connected to the uncut ends of the stock material. Once all operations are complete, the supports are severed (sawed or milled) in a final series of operations and the part is removed. Post-processing is performed to finish the minimal support contact patches on the part.

The setup and steps to this approach are shown in Figures 7-2 and 7-3, respectively. As an example in Figure 7-3, a component is being machined using sacrificial supports to retain the part at its ends along the axis of rotation. We have employed this setup and process planning strategy in previous work as a method for rapid machining. Sacrificial support fixtures have been designed manually and used in several proof-of-concept parts. This method of using one axis of rotation for indexing between setups is obviously not capable of machining *all* parts of extremely complex shape. Parts with severely undercut features or complex features on three or more mutually orthogonal faces may not be machinable with this approach. In particular, this setup strategy assumes that *some* axis of rotation exists such that all surfaces are visible.

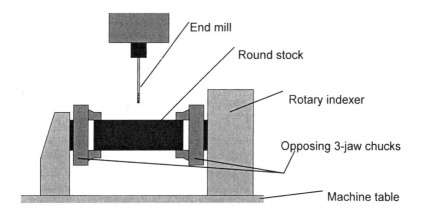

Figure 7-2. Setup for CNC-RP

In the case where a single axis of rotation is used for setups, we have developed a visibility method that can determine if an axis of rotation is feasible, and determine the minimum set of rotations such that all surfaces are visible[13]. More recently, a method for calculating machinability using only slice geometry information has been developed[14]. This latest work has yielded a set of algorithms that can determine if a particular tool can access the surfaces of the part without the use of feature recognition approaches.

The methods developed in this chapter trade off some additional machining time in order to eliminate a significant set up time for part offset location and fixturing. Our preliminary results from initial experiments with this approach are extremely promising for not only functional prototypes and very small quantities, but also for small batch production where setup for offsets is a significant task.

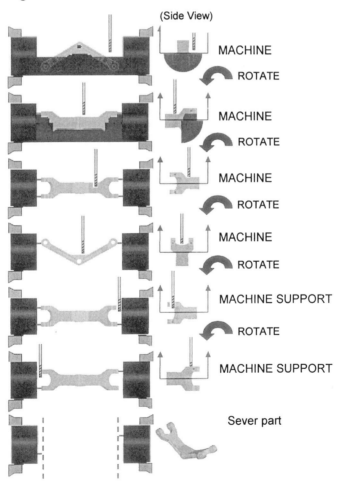

Figure 7-3. Process steps for CNC-RP

A preliminary example of a part created using the CNC-RP methodology is shown in Figure 7-4 where a functional suspension component for a mountain bike was created from a section of round stock material (Steel). In this case, four sacrificial supports were used, two on each end of the part along the axis of rotation. These supports were designed and added to the CAD drawing manually and with an arbitrarily small diameter. Upon completion of the part geometry, two of the supports (temporary supports) were removed in a final machining operation (Figure 7-4a). Lastly, the part was cut from the stock by sawing the two remaining supports (Figure 7-4b). This relatively complex part would have been a formidable challenge using any of the traditional methods of fixturing. Also, maintaining offset locations during each removal, rotation and re-clamping for each orientation of the setup using existing methods would have been a considerable problem.

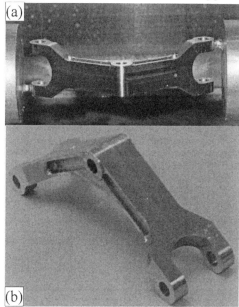

Figure 7-4. A sample part: a) in process using sacrificial supports and b) upon

7.5. Approach to Setup Planning

The challenges in creating *layer-based* toolpaths are not in the actual cutter location data generation. A commercial CAM software package such as MasterCAM can easily generate 2½-D toolpaths. It is as simple as setting the *maximum step down* parameter to the desired layer depth. The critical steps in the toolpath planning method are to determine; 1) How many orientations about the axis are needed in order to machine the part?, and 2) Where are they? The problem is two-fold: 1) determine whether all surfaces of the model are machinable with rotations about the selected axis, and if so,

2) calculate the minimum number of orientations (index rotations) required to machine the entire surface. A *necessary* condition for a surface to be machinable is that it must at least be visible. Other *sufficiency* conditions exist including tool reach and proper cutter contact for complex surfaces. For example, a ball-end mill will need to have sufficient length to reach a surface and be able to contact the surface with some point on the hemispherical end of the tool.

We address the necessary condition of *visibility* using a simplified approach that does not require feature recognition. Since tool access is restricted to directions orthogonal to the rotation axis, 2D visibility *maps* for a set of cross sections of the surface of the model are used for finding the set of orientations for machining. This procedure approximates visibility to the entire surface of the model. For example, consider the part illustrated in Figure 7-5. Cross sectional slices of the geometry from an STL model provide polygonal chains that are used for 2D

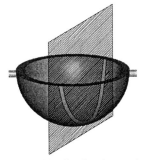

(a) Model is sliced orthogonal to the rotation axis

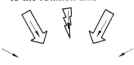

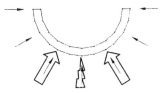

(b) Tool access directions are restricted to the slice plane

Figure 7-5. Model with sample cross section used for visibility mapping

visibility mapping. A simultaneous visibility solution for all cross sections of the model will approximate visibility of the entire surface. For this simple model and the slice shown in Figure 7-5, the chain of edges in the polygon can be "seen" from many different views. If the views in Figure 7-5 illustrated by the block arrows are chosen (⇒), four rotations *could* be used to machine the part. This implies that four orientations (index rotations) are used and all visible material from each view is removed. If the two orientations noted by the lightning arrows (⚡) are used, then only two rotations are needed. In this case, two rotations is the fewest number required. The entire presentation of the visibility algorithms will not be covered in this chapter; however, a general description of the approach to visibility is provided.

Visibility for each polygonal chain is determined by calculating the polar angle range that each segment of the chain can be seen (Figure 7-6). Since there can be multiple chains on each slice, one must consider the visibility blocked by all other chains. Therefore, the visibility data for each segment can be a set of ranges (Figure 7-6). If a visible range exists for every segment on each chain, for all slices in the set, then the remaining problem is to determine the minimum set of polar directions such that every segment is visible in at least one direction. The problem of finding the minimum set of rotations sufficient to see every surface of the model can be formulated as a *Minimum Set Cover* problem. One will note that the *Minimum Set Cover* problem is NP-Hard. Since large instances of NP-Hard problems do not have known solutions that can be solved consistently or efficiently[15], an approximate solution is found using a *greedy* approach[16], employed after the visibility mapping is complete. The solution of the set cover provides the minimum set of angles from the set [0°, 360°] such that, for every segment, at least one angle is contained in one of its visibility ranges.

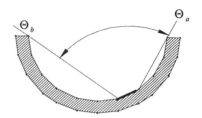

(a) Visibility for the segment= $[\Theta_a, \Theta_b]$

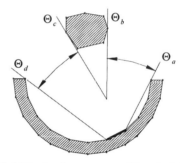

(b) Visibility for the segment= $[\Theta_a, \Theta_b], [\Theta_c, \Theta_d]$

Figure 7-6. Visible ranges of one segment of a polygonal chain

Visibility analysis can be used early in the analysis of the part geometry, but before toolpaths are generated, we need to determine if the surfaces of the part can be accessed by the tool. Once the system has determined the tool(s) that will be used, the *machinability* analysis algorithms can be initiated. The machinability analysis approach used in CNC RP is based on the concept of Configuration Space (C-space). Due to space requirements, it will not be described in detail in this chapter. The basic idea of C-space is to find the aggregate of the valid spatial configurations for a moving mechanism in an environment with obstacles around it (For CNC RP: A tool, with the segments from slice cross sections as obstacles). C-space of the cutting tool, which we define as *Tool-Space*, provides the safe space for tool path planning, therefore tool paths based on C-space are always gouge-free and collision-free. If Tool-Space exists, then at least one tool path can be generated to machine the corresponding geometric point on the part surface, and hence this point has machinability. Therefore, by evaluating the Tool-Space of each point on a surface, the

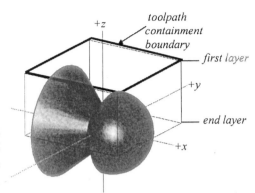

Figure 7-7. Layer-based toolpath

machinability of this surface can be determined. Therefore, we can determine if a set of orientations that provides visibility actually provides machinability, depending on the tool chosen.

Once the set of orientations is determined, each orientation's toolpaths can be generated. Since this is a feature-free approach, the selection of surfaces for each machining operation is straightforward; **all** surfaces of the part are used for toolpath planning for **every** orientation of the solution set. Also for each orientation, the containment boundary for creating layer-based toolpaths must be defined. Assuming the tool is oriented in the z-direction, the containment boundary above the part is specified by a rectangle (x-y). The other information required is the depth of the maximum and minimum z-level layers (see Figure 7-7). The length of the boundary (x) must be greater than the part length along the axis of rotation, while the width of the boundary (y) must be greater than the maximum part diameter. Specifically, the containment boundary must be greater in both length and width of the part by at least the diameter of the tool (on all four sides). This is necessary since the tool requires a path around the part equal to its diameter, in order to machine around the visible boundaries of the part.

The toolpath layers must begin at or just above the stock surface and proceed through the distance (z) to the furthest visible surface from the current orientation. To find this furthest distance, we can calculate the maximum distance to all segment endpoints

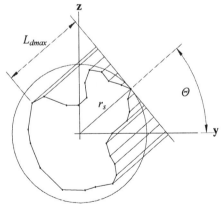

Figure 7-8. Distance to the deepest visible segment at one orientation

visible from each orientation in the solution set from the visibility algorithms. The data from the visibility method provide the set of segments visible from each orientation in the solution set. Each segment is defined by its endpoints (P_i, P_{i+1}) where each endpoint has coordinates in the y-z plane (y_i, z_i). The perpendicular distance from each point to the tangent line at the solution orientation is calculated. The maximum distance to all points visible from each orientation is used as the location of the *maximum layer depth* for that orientation (Figure 7-8).

Although the STL representation will be used for visibility mapping, if a solid model exists, then it is used for toolpath planning in CAM. The model is simply rotated in the CAM environment to each of the orientations from the visibility algorithm and the toolpaths are created using the other setup parameters from the slice file information.

The setup planning approach described relies on an axis of rotation. It is therefore important to determine as early as possible if a chosen axis is feasible. Moreover, if one is chosen that is not feasible, then the user must be clearly prompted the surface areas that will not be possible using the selected axis. In future work, we will create a method for determining necessarily infeasible axes. These would be axes that occlude visibility to some surfaces regardless of the potential diameter or length of the tool available. A goal is to provide graphical guides to aid the user in selecting an axis. However, selecting an "optimal" axis will be a challenging problem to address in later research. Axis selection not only needs to ensure complete machinability, but should do so in a manner that minimizes the number of setups and overall machining time. Axis selection will impact machinability directly, if for example, the stock diameter becomes large due to a poor axis choice. This will drive up tool diameters and therefore will likely drive down tool accessibility.

7.6. Approach to Tool Selection

Proper tool selection must ensure collision free machining for any model complexity. One condition to ensure collision free toolpaths is that the tool length must be greater than or equal to the distance to the furthest visible surface with respect to current orientation. In this manner, one is assured that even on the deepest layer, the tool holder will not collide with the stock (see Figure 7-9).

To ensure that no portion of the tool will collide with any previously machined layers, the tool shank diameter must be less than or equal to the flute diameter.

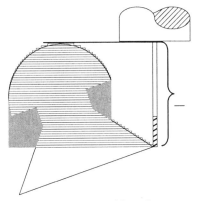

Figure 7-9. Tool length requirement

This criterion unfortunately makes a tool more susceptible to deflection and breakage. Typically, long tools are designed with large shank diameters and only have the length of the cutting surface (flutes) at the prescribed diameter. Figure 7-10 illustrates a tool with the proposed characteristics reaching to a z-depth without tool collision.

A desired goal is to choose tools that will be capable of machining a variety of complex surfaces at the required accuracy. The current approach is to select the smallest tool diameter available in the necessary length that is specified.

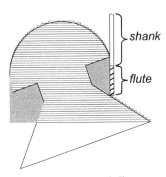

Figure 7-10. Tool diameter requirement

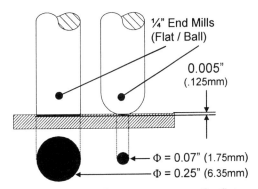

Figure 7-11. Cutter contact area for flat- and ball-end mill

Since only 2½-D layers are to be machined, a flat-end tool is most appropriate. Whereas a ball-end (spherical) tool is able to machine smaller radii surfaces in some cases, the diminishing diameter of the cutter contact patch is a problem since very shallow depths of cut are used each layer (see Figure 7-11).

As noted, one of the goals in tool selection for rapid CNC is to minimize the cutter diameter. This is directly opposite of the goals of a typical manufacturing process plan. However, based on the assumption that feature information is not known, one must use the smallest diameter tool available in a given length. From a purely geometric standpoint, this increases the likelihood that the smallest features of the part can be machined. Tool selection is both related to, and impacts, other process planning parameters. For example, the diameter of the tool defines the extent of the toolpaths along the rotation axis direction. This affects the length of the sacrificial support cylinders since they need to protrude from the ends of the part.

7.7. Challenges with Rapid Fixturing

Designing a fixture scheme for CNC machining is a difficult task that requires a significant amount of work from a highly skilled technician. In general, fixturing or *workholding* serves three primary functions: location, clamping, and support[17]. Just as the approach to developing toolpaths of this research differs from traditional machining, the fixturing requirements for rapid machining are significantly different. Typically, a human operator orients the stock material between setups. Fixtures, in combination with hard stops and/or probes, are used to establish reference locations on the stock material. If a part is to be machined in multiple setups, it is critical that the fixture scheme facilitates repositioning of the stock such that dimensional constraints can be satisfied. Once the stock is located in the fixture, sufficient clamping force is needed to withstand the machining forces. In addition, the fixture must provide support for overhanging or slender features so that the stock material will not deflect too much under the machining forces.

The goals of rapid machining make the fixturing problem both more and less difficult. In rapid machining, accessibility to the surfaces of the part is of paramount importance. If we were to use a traditional parallel jaw vice, for example, much of the part surfaces would be contact with either the jaws or the bottom of the vice. This is not always a problem in traditional CNC machining, where a piece of stock (usually prismatic in shape) is *populated* with features. For example, in one setup, the process plans for a part may include a pocketing operation, then a slotting operation in the bottom of the pocket. The face of the stock where these features are being created may be cut during the machining operation, or created in a preprocessing step, or it may simply be the result of the original stock production. In the current method, one does not assume that any shaping operations have occurred nor is the original stock shape considered viable for a finished feature. The current methodology suggests a *feature-free* manufacturing approach whereby all surfaces of the part are candidate "features" to be machined from an orientation. Consider a comparison between a typical machining approach and the current methodology, as illustrated with a simple part in Figure 7-12. In Figure 7-12b, the pocket and slot are machined on the face of a block, with the sides of the block clamped by the jaws of a vice. Since these are the only two features to be machined in this setup, then the vice fixture is appropriate. This is the case where the stock material is being populated with two features, namely the pocket and slot. In Figure 7-12c, the same part is being machined out of a larger piece of stock material. In this case, the top, front, back and sides of the block must be machined (to some depth) in addition to the pocket and slot on the top.

Recall, that for each setup orientation, **all** surfaces are used in process planning. The intent is that all visible surfaces from any orientation *may* be machined in that orientation. The feature-free nature of this method demands that the fixture solution provide as much access to the part as possible. Each rotation places the stock material in a new *setup* orientation; however, the work offset must be retained for each orientation without having the user re-establish it. Therefore, the fixture solution for rapid machining must allow rotations of the part without the need to relocate reference points. The current fixturing method takes advantage of the fact that reorientation of the stock is only executed about one axis, which makes it easier to develop an automated method.

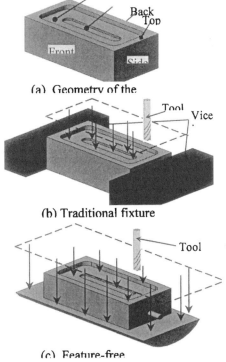

Figure 7-12. Comparison of traditional

Typical cutting forces during machining can cause the stock material to slide or shift and some elements of the part can deflect under load. A characteristic of the rapid machining method is that cutting forces are significantly lower in magnitude (similar to high speed machining) and less variable since the depth of cut is very shallow and is the same for all operations. This results in a significant decrease in the amount of clamping and support that the fixtures need to provide. The cutting forces for CNC RP are orders of magnitude less than those of a typical machining process plan.

A significant challenge for developing a fixturing scheme for rapid machining is that the fixtures must be generated automatically. Our approach to fixturing for CNC RP borrows from the general idea of *sacrificial supports*, which are used in existing RP systems. In this work the general intention is retained; however, the requirements for the support structures are different. The goal is to have a fixture solution that is created in-process and is customized for each part. Specific to this work, the fixture supports need to allow the part to be rotated about the axis while providing access to as much of the part surface as possible. Conventional fixturing methods for CNC often utilize vices, clamps, vacuum surfaces, etc. These approaches occlude visibility to a significant amount of the part or make it difficult to reorient the part for multiple setups.

In the CNC-RP method, the *sacrificial supports* are added to the ends of the model (in CAD) such that the model remains attached to the round stock material throughout the set of machining operations. In the current implementation of the method, small diameter cylinders are manually added to the CAD model at its ends prior to creating the process plans, so that toolpaths are created to machine the cylinders in the same layer-based fashion that the model is processed. At least one sacrificial support is necessary, but numerous ones may be required to fixture the part during machining operations. This concept was illustrated in Figure 7-13 where a finished part is fixed within a cylindrical piece of stock material, which in turn is fixed between the jaws of two opposing chucks. For every orientation about the axis, the relative tool (z=0) offset location is constant at the center of rotation. Similarly, the part coordinates remain at a consistent location with respect to the stock material clamped in the vice and located on the table for every rotated orientation.

An advantage of using sacrificial support fixturing is that, for a given setup orientation, the amount of visible surfaces can be increased compared to traditional fixtures. If a traditional vice were used and features needed to be machined on numerous surfaces, the part will need to be unclamped, reoriented and then re-clamped for each orientation. It is difficult to reorient and re-clamp a part without introducing error during location. Utilizing sacrificial supports, the stock is reoriented without changing the relative location of the part with respect to the machine table. Once the complete set of rotated toolpaths has been executed, the supports are cut by the user, which releases the part from the round stock material. This of course adds a post-processing step, where the surfaces of the model at the contact point of the cylinders must be sanded, ground, etc. This has been improved by generating a final set of machining operations that focus on reducing the cross section of the final two supports for easy removal.

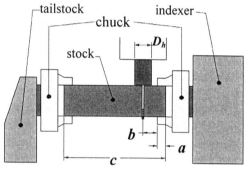

Figure 7-13. Setup for Rapid Machining

Stock length: $c = L_p + 2a + 2b$
where: L_p = Part length
a = Clamping length
b = Collision offset $(x) = .5D_h + .5D_t max$
D_h = Diameter of tool holder
$D_t max$ = Diameter of largest tool

In addition to the technical advantages of using sacrificial supports, there are practical advantages with respect to making this rapid machining method straightforward to implement. To setup the workpiece for the current method, the user clamps a piece of round stock between two chucks. The diameter of the stock is simply as large as the diameter of the part about the rotation axis, and its length can be calculated in a straightforward manner (Figure 7-13). The *collision offset* (b) ensures that a part program can be run with no risk of collision between the tool holder and either of the chucks. It is assumed that the rest of the spindle will not collide, given a proper choice of tool holder length. A significant advantage of this fixturing methodology is that work and tool offsets do not need to be set before running a part program. The *collision offset* (b) can be used to translate the part coordinates such that tool collision cannot occur, given a tool holder, and maximum tool diameter. This makes setting the work offsets for each new part unnecessary. The work offset coordinates [x(0), y(0)] will always be aligned with the axis and located at the face of the stationary (along x-axis) chuck on the indexer. Similarly, the tool offset is set to the z-height corresponding to the axis of rotation, as mentioned previously. If a proper length and diameter stock is clamped between the chucks, a part program can be executed collision-free with no need for the user to set offsets; a time consuming and often error-prone task in CNC machining.

Sacrificial fixturing is currently a manual activity, although general design guidelines are being developed. In our current research, the sacrificial support structures employed are designed based on simple cylindrical structures. In future research, we will fully integrate sacrificial support design and execution with the process plans. Mechanics and Dynamics models will be created that approximate deflections of the model and support structure skeleton. The sacrificial support design methodologies will also be tied into the visibility and machinability analysis since support structures, both temporary and permanent, will affect the accessibility of tools during the sequence of machining operations.

7.8. General System Model

The diagram in Figure 7-14 represents some of the current and future components envisaged for the CNC RP system. The most critical interactions and data transfer is represented; however, there are numerous connections that still need to be mapped.

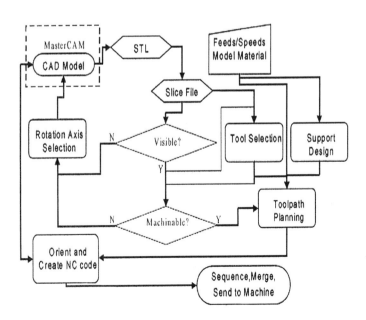

Figure 14 – CNC RP System Model

7.9. Example Parts using CNC-RP

The visibility algorithms were implemented in C and tested on a Pentium IV, 2.0Ghz PC, running Windows XP. The software accepts slice files as input and returns several critical process parameters; 1) The minimum number of orientations, 2) The minimum stock diameter, and 3) The distance to the minimum and maximum layer depth for each orientation.

Using the set of orientations, max/min depths of cut, and stock diameter, toolpath plans can be generated using CAM software. Several prototypes have been machined in the laboratory. Although the intent is to integrate the visibility software with CAM and automate process planning tasks, at present the steps of toolpath processing are done manually. The steps that were executed are as follows:

1. Visibility software is executed.
2. CAD model is rotated through each of the orientations of the visibility solution.
3. Toolpath containment boundary created using stock and tool diameter and length of the part.
4. For each orientation, *rough surface pocket* toolpaths (MasterCAM) are generated. Minimum depth set at stock radius and max depth set to the parameters given by visibility software.
5. NC code for each orientation combined manually into a file with 4th axis rotation commands.

Time required for step 1 is on the order of seconds, and less than ½ minute for most parts. Steps 2-5 require 5-15 minutes depending on the number of rotations and processor speed of the computer. The following are example prototypes machined in the laboratory.

The first prototype is the toy Jack. The Jack was machined on a Haas VF-0 3-axis machining center. The number of orientations provided by the visibility method was five. The

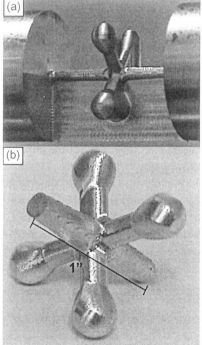

Figure 7-15. Jack model (a) Between orientations and (b) Finished part

part was created in approximately 3 hours. Figure 7-15a shows the prototype of the Jack in between machining operations while 7-15b shows the jack after being cut from the stock at the sacrificial supports once all orientations were machined.

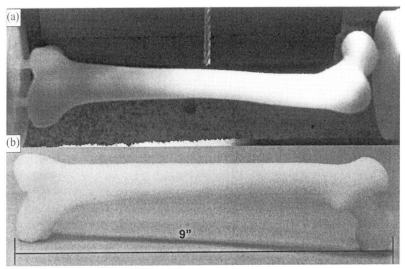

Figure 7-16. Bone model (a) Attached via sacrificial supports and (b) Finished model

The next model is of a human leg bone; the femur, which was machined from Delrin plastic. Figure 7-16a presents a view of the femur prototype during processing. As can be seen, the bone model is secured to the remaining stock via three sacrificial supports. The stock material is clamped on both ends in 3-jaw chucks, one on a tailstock and one on the face of a rotary indexer. Figure 7-16b shows the finished prototype after machining from 3 orientations. The 3 orientations for machining are illustrated by the arrows in Figure 7-17, as viewed from the distal end of the femur (left end in Figures 7-16a,b). Total machining time was approximately 12 hours.

The final prototype example is a simple prismatic part, a block with 3 through-holes. Although this is a considerably simple part to machine using traditional methods, it can easily be measured on a coordinate measuring machine (CMM) in order to evaluate the accuracy of the current process. (Figure 7-18). Again, the part was made without the use of a tailstock. A grossly oversized sacrificial support was used on one end to ensure stiffness for this test.

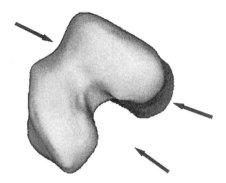

Figure 7-17 – Visibility orientations for the femur

The prismatic block was measured on a Zeiss Vista CNC CMM. A runout error of 0.002"(50microns) was detected in the fixtured stock prior to machining and is presumed to be the source of much of the measured error, in particular, the undersizing of the width of the part. Overall the largest deviation in dimensions was on the order of 0.005"(125microns), but it is expected that machine accuracy can be achieved with a fully implemented fixture scheme. In particular, runout with respect to the axis of rotation should be eliminated using the sacrificial fixturing method using opposing 3-jaw chucks.

The same prismatic part was also created using a Sterolithograpy (SLA) machine (see Figure 7-19). The processing time on the SLA machine was estimated using a software build time estimator[18], at 2 hours and 56 minutes; however, the laser on the machine was old (laser power reduced from 35mW to 19mW) and the part required additional time (total of 4 hours and 46 minutes).

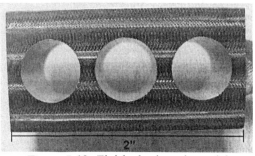

Figure 7-18. Finished prismatic model

For the SLA part, the larger deviations in dimensions were the diameter of the holes (0.004"- 0.005"/100 –125 microns) and the largest was in the height of the part (~ 0.02"/500microns). The total time to create the SLA part was ~ 7 hours. This included 4 hours and 46 minutes to build it on the machine, 15 minutes to clean it and remove the supports and finally, 2 hours in the post-curing oven.

Figure 7-19. SLA model

Although the CNC machined prismatic block was not built with two appropriate sacrificial supports, it is estimated that removal of the supports would have required ~15 minutes, as was true with the Jack and the Bone. A comparison of the actual, or estimated[18,19], build times for creating the three parts using rapid machining (CNC RP), Fused Deposition Modeling (FDM) and SLA are shown in Tables 7-1 and 7-2.

For the three example processes, rapid prototyping using CNC machining is shown to be the fastest of the three rapid methods in all but one case (vs. SLA of bone). There is also the added benefit of having a part made of better materials (Aluminum, and Delrin plastic) rather than ABS plastic or photosensitive resin.

Table 7-1. Build and Post Process Times

	Estimated build time			Actual build time			Estimated post process time		
Process	block	jack	bone	block	jack	bone	block	jack	bone
CNC RP	*	*	*	3h 30m	3h	12h	15m	15m	15m
SLA	2h 56m	3h 56m	9h 34m	4h 46m	*	*	2h 15m	2h 15m	2h 15m
FDM	4h 16m	1h 31m	15h 48m	*	*	*	2h	2h	2h

Table 7-2. Comparison of Total Processing Time

	Total processing time (est. and/or actual)		
Process	block	jack	bone
CNC RP	3h 45m	3h 15m	12h 15m
SLA	7h 15m	6h 11m	11h 49m
FDM	6h 16m	3h 31m	17h 48m

7.10. Economics of CNC-RP

The CNC-RP process can be a very cost effective method for creating functional parts and can also be used for prototyping. The intended use of the process is for accurate parts in the correct materials with good properties. In other words, one would expect to use CNC-RP if the parts created will be physically tested. If the intent is to simply create a model for passing around a meeting table, then CNC-RP may not be a wise choice. As mentioned previously, there are concept modelers available that can create models of CAD designs very quickly, although they use plastics and even cornstarch powder to create them. So, the parts may not be very strong, but they can be created quickly and cheaply on a ~$25K machine. The CNC-RP process may not be a good choice for this type of model because the tooling and material costs will likely be higher. Also, the base price on an industrial milling machine is about $50K and requires more overhead in electrical, air, coolant and technical support.

However, if the part needs to be made from a stronger plastic that is available on an existing RP machine and the accuracy is within the capability of an existing RP machine, then the decision to use CNC-RP is less clear. The mid range RP machines typically cost in the range of $150K-$300K and have tolerances as good as +/- 0.010" and use more common plastics such as ABS and Polycarbonate. In this case, the economics may come down to a comparison of the volume of material that would need to be ADDED using a current RP process versus the volume of material that would need to be SUBTRACTED in a CNC-RP process. Consider the two parts shown in Figure 7-20 below. Figure 7-20a shows a part with a very small mass/enclosing volume as compared to the part shown in Figure 7-20b. The part in Figure 7-20a would require an excessive amount of machining to remove the material inside, however, this would be reasonably efficient for an additive RP process (even with the support structure in some processes). However, for the part in Figure 7-20b, an additive RP process would spend an excessive amount of time stacking simple layers, whereas the milling process would finish the part very quickly from a block of material. The economics of using CNC-RP in this case would likely depend on the material costs; assuming that the same materials could be used in each case and tolerances were capable of being held using the additive RP process.

We do not discuss the cost comparison of using CNC-RP with other RP processes for creating *functional* parts since this is rarely if at all possible with the existing processes. In the previous section we discuss a scenario where the material and tolerance requirements are not stringent, so one might consider using an additive versus subtractive process. However, if the part needs to created from a material like steel and has tolerances less than a few thousandths of an inch, then rapid machining is the clear choice because no current RP process has this capability.

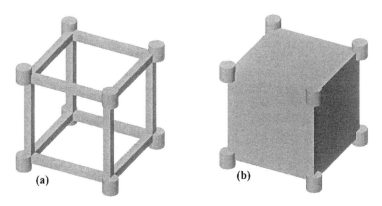

Figure 7-20. Example models with low (a) and high (b) mass/enclosed volume

Lastly, we must consider the use of the CNC-RP process for creating functional parts instead of using traditional production manufacturing approaches. This might mean using CNC-RP for custom designed products or perhaps for creating replacement/service parts. For example, if we needed to replace a part from a military vehicle that is decades old, then it may not make sense to use the original process steps for production. This might have included creating a pattern/mold for casting a near-net-shape part, then machining the critical features using a series of fixture setups. This may not make sense if all we really need is one part; therefore, the CNC-RP process might be a better choice. The major driver behind using the CNC-RP process when compared to traditional methods is likely the number of setups required. For example, if a part is very simple and can be machined in one setup from a block of material in a traditional vice, then CNC RP makes no sense. However, if multiple setups are required, then a sacrificial fixturing process like we describe may be a better approach.

In a traditional machining process, one would need a very specific set of plans for flipping a block of material between operations and orienting it for each operation. This is not very difficult if the features are machinable from the three orthogonal orientations (features located on the faces of a cube/prismatic block, for example). However, if some feature needs to be created in an orientation that is not parallel to one of these orientations, then the fixturing problem can be very complex. Consider, for example, the part illustrated in Figure 7-21. This part is a simple cube with an extruded-cut feature that passes through two faces. In order to create this feature, the block would have to be oriented such that the feature is parallel to the tool axis. This would require a more complex fixture that can clamp the part in the correct orientation and maintain the reference coordinates of the part. Now, consider a part that has two, three or several features such as this; the fixturing complexity will grow exponentially. Moreover, the accuracy of the part will likely suffer since each setup will have some error in location and orientation. So, the question is: How complex of a part does the CNC-RP process become viable/economical?

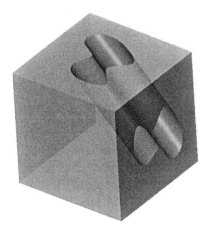

Figure 7-21. Example part with non-orthogonal feature

The answer is unfortunately not straightforward; however, one can easily see where the sacrificial fixturing approach provides several advantages. Even without knowing the cost of each fixture setup, we know that the CNC-RP process has no real cost associated with rotating the part to a new orientation (a simple rotation of the indexer). The reality is that the number of rotations simply means more machining time since more orientations of layer-based removal must take place. On the other hand, each rotation does have a direct cost in a traditional machining operation where the setup planning and qualification of each fixture setup would be very costly if the number of setups is large.

It is of course difficult to clearly describe the conditions where CNC-RP is a viable approach to creating a part. However, it can be generally stated that the materials and accuracy requirements for a model will affect the decision of whether to use an additive or subtractive process. Furthermore, it is clear that the number of parts to be produced and the complexity of the part are the critical factors in choosing between CNC-RP and traditional machining and fixturing methods. For a very small batch of complex parts where the material properties and tolerances are critically important, it is probable that CNC-RP is the most appropriate method and in some cases, the only method.

Lastly, one might simply consider the processing times for CNC-RP versus additive methods and traditional machining approaches. The preliminary data (Tables 1 and 2) shows that the method is currently comparable to additive methods. However, we have not even broached the topic of optimization of this process. It should be apparent how much "air cutting" CNC-RP does during its layer-based removal processes, in particular when machining from several orientations. We currently have not created the necessary modifications to the toolpath planning strategy so that previously machined surfaces in one orientation are not machined again in another. Conservative estimates from our preliminary experiments suggest that over 50% of our processing time is spent machining the air. In addition, we are using a 7500 RPM spindle on an entry-level machine; therefore an additional drastic reduction in processing time could be realized by simply using High Speed Machining (HSM). We are currently machining with very small depths of cut at high feed rates, so the progression to a high speed machining approach should be straightforward. We envisage that the optimized system using high speed machining will be able to reduce processing times of metal parts to just a few hours, from CAD to Part. When the processing times are reduced significantly, the CNC-RP process may prove to be the best choice for custom machining or very small batch production.

7.11. Limitations and Future Work

This CNC-RP process represents the first system that is capable of creating models in a truly wide array of materials with very high accuracy. With any system there are obviously limitations in what can be done. This process is basically an automated machining system, so of course much of the existing limitations in any machining process persist. The method can be used for moderately complex part geometries. That is, parts with complex geometries that are accessible by rotations about one axis are possible, however, even simple geometries that are not visible about the axis of rotation are not machinable. Parts with severely undercut features can also be a problem and hollow parts are of course impossible. In addition, small inside corner radii are difficult or impossible to machine, depending on tool geometry. Not all parts will have a feasible axis of rotation such that all surfaces can be machined using the current CNC-RP method. An example of a prismatic part that would not have a feasible axis of rotation is shown in Figure 7-22a. This prismatic block has three pockets located on mutually orthogonal faces. As such, at most two pockets could be machined from one axis of rotation. The next example is a spherical shaped part with several slots about its surface (Figure 7-22b). If an axis is chosen such that all slots can be machined (*A1* in Figure 7-22b), then a significant amount of the interior will not be accessible. If an axis is chosen such that the entire interior can be milled (*A2* in Figure 7-22b), then only as many as two of the slots can be completely machined.

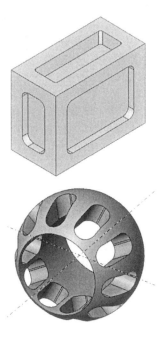

Figure 7-22. Parts with no feasible axis

The visibility algorithms answer the question of whether an axis of rotation is feasible but currently do not search for a better solution. In its current implementation, the visibility method accepts a model oriented by the user and generates visibility information based on that axis of rotation. The current research is focused on developing a method for evaluating multiple orientations and guiding the user at least semi-automatically to a "better" axis of rotation. In this case, "better" can mean several things; 1) less material, 2) shorter tools, 3) fewer rotations, to name a few. In addition, we are currently creating a method for evaluating the "machinability" of a part model. This means that we are trying to take into account the size and shape of the cutting tool in deciding the set of orientations for machining. Since visibility is only an approximation of tool accessibility, then we do not know if the entire part surface is "machinable" using the set of orientations from the visibility map. The machinability analysis being developed is based on flat end milling with a simple tool of equal flute and shank diameter. We simply input the tool diameter to the analysis and then map the machinability of the part surfaces. Similar to the visibility analysis, we only need to study the 2D slices of the part to evaluate machinability about the rotation axis.

Lastly, a significant research effort is being directed at the automatic generation of sacrificial support fixturing for CNC machining. Sacrificial supports will greatly reduce the cost in both prototyping using CNC machining, and in many cases in short production runs or batch processing of parts.

7.12. Summary

This chapter presented a new methodology for using CNC milling as a rapid prototyping process. The method makes it possible to rapidly plan and create machined parts and prototypes with little or no human intervention. The method presented is a completely "feature-free" approach to machining and is capable of analyzing any part geometry and creating a set of plans for machining the part automatically. This does not require any user-input to guide the process, as would be true in a typical planning process using a skilled machinist. As we move forward, digitized information for CAD files is becoming more popular and freeform shapes are more commonplace in 3D CAD design. Today, parts are being designed with complex shapes, either for aesthetic, functional or ergonomic reasons and therefore feature recognition will be practically impossible. As an example, a part like the human femur bone has no defined features such as holes, slots, or pockets. Visibility and machinability analysis using 2D slice geometry has made it simple to extract critical process planning information. The research has also further developed the concept of *sacrificial supports* for use in a subtractive process. Sacrificial supports have been used extensively in additive RP and in an ad-hoc manner in machining processes. However, no previous or current work cites the formal development of design and execution strategies for sacrificial support fixturing used in an automated machining process. This research lays the initial groundwork for future work in sacrificial fixture development for machining. The combination of the process and fixture planning methods outlined in this chapter makes the concept of CNC-RP a significant step toward a completely automated rapid machining process.

Reference:

1. P.G Maropoulos, Review of research in tooling technology, process modeling and Process planning, Part 2: Process Planning, Computer Integrated Manufacturing Systems, **8**(1), 13-20 (1995).
2. Y.H. Chen,. Y.S. Lee,.and S.C. Fang, Optimal cutter selection and machining plan determination for process planning and NC machining of complex surfaces, Journal of Manufacturing Systems, **17**(5), 371-388 (1998).
3. A. Joneja and T.C. Chang, Setup and fixture planning in automated process planning systems, IIE Transactions, **31**(7), 653-665 (1999).
4. Z.M. Bi, and W.J. Zhang, Flexible fixture design and automation: Review, issues and future directions, International Journal of Production Research, **39**(13), 2867-2894 (2001).
5. Y.H. Chen, and Y. Song., The development of a layer based machining system, Computer Aided Design, **33**(4), 331-342 (2001).
6. R. Merz, F.B. Prinz, K. Ramaswami, M. Terk, and L.E. Weiss, Shape Deposition Manufacturing, Proceedings of the Solid Freeform Fabrication Symposium, University of Texas at Austin, 1-8 (1994).
7. S.E. Sarma, and P.K. Wright, Reference Free Part Encapsulation: A New Universal Fixturing Concept, Journal of Manufacturing Systems, **16**(1), 35-47(1997).
8. B.S. Shin, D.Y. Yang,. D.S. Choi, Lee, E.S., Je, T.J., and Whang, K.H., Development of rapid manufacturing process by

high-speed machining with automatic fixturing", Journal of Materials Processing Technology, **130**(1), 363-371 (2002).
9. D.S. Choi,., Lee, S.H., Shin, B.S., Whang, K.H., Yoon, K.K., and Sarma, S.E., A new rapid prototyping system using universal automated fixturing with feature-based CAD/CAM, Journal of Materials Processing Technology, **113**(1-3), 285-290(2001).
10. E. Radstok, Rapid Tooling", Journal of Rapid Prototyping, **5**(4), 164-168 (1999).
11. F.A. Vouzelaud,., Bagchi, A., and Sferro, P.F., Adaptive Laminated Machining for Prototyping of Dies and Molds, Proceedings of the 3rd Solid Freeform Fabrication Symposium, 291-300 (1992).
12. D.F. Walczyk,., and Hardt, D.E., Rapid tooling for sheet metal forming using profiled edge laminations-design principles and demonstration, Journal of Manufacturing Science and Engineering, Transactions of the ASME, **120**(4), 746-754 (1998).
13. M.C. Frank, Wysk, R.A., and Joshi, S.B., Determining Setup Orientations from the Visibility of Slice Geometry for Rapid CNC Machining, ASME Journal of Manufacturing Science and Engineering, (*accepted, to appear*)
14. Y. LI, and Frank, M.C., Machinability analysis for 3-axis flat end milling", ASME Journal of Manufacturing Science and Engineering, (*accepted, to appear*)
15. C.A., Tovey, Tutorial on Computational Complexity, Interfaces, **32**(3), 30-61 (2002).
16. V. Chvátal, A greedy heuristic for the set-covering problem, Mathematics of Operations Research, **4**(3), 233-235 (1979).
17. T.C. Chang,., Wysk, R.A., and Wang, H.P., *Computer-Aided Manufacturing*, Prentice Hall Inc., New Jersey (1998).
18. Georgia Institute of Technology, Rapid Prototyping and Manufacturing Institute, *Build Time Estimator, Version 1.2*, Copyright, Georgia Tech Research Corporation, 2002, Url:http://www.rpmi.marc.gatech.edu/, Accessed: July, 2003
19. Stratasys, Inc. 14950 Martin Drive, Eden Prairie, MN 55344-2020 USA, Url:http://www.stratasys.com/, Accessed: June, 2003

Chapter 8

SELECTIVE INHIBITION OF SINTERING

Behrokh Khoshnevis[1] and Bahram Asiabanpour[2]
[1]*University of Southern California, Los Angeles, CA 90089-0193, khoshnev@usc.edu;* [2]*Texas State University-San Marcos, TX 78666, ba13@txstate.edu*

Abstract:
SIS is a new method of building 3D objects using powder sintering. The method is capable of making plastic parts without the use of laser. An Alpha machine has been constructed and preliminary studies have been carried out to prove the concept and to explore the process variables and their impacts. Various polymers including polystyrene and polyester have been successfully used in our experiments. The approach should be feasible for a variety of polymers.

Key words:
Layered Fabrication, Sintering, Sintering Inhibition, Droplet Deposition, Rapid Prototyping, SIS, NC Path Generation, Material Selection

8.1. INTRODUCTION

The Selective Inhibition of Sintering process is a new RP method that, like many other RP processes, builds parts in a layer-by-layer fabrication basis. The SIS process works by joining powder particles through sintering in the part's body, and by sintering inhibition at the part boundary. As shown in Figure 8-1, the SIS process starts by laying a thin layer of powder slightly above the previous layer, by sweeping a roller over both a powder supply tank and the build tank (A). Then, the areas of the powder bed selected for sintering inhibition are wetted by a printer (B). A radiation-minimizing frame is positioned to prevent areas of the powder layer which lie outside the part envelope from sintering (C). The entire layer is then sintered with a blast of thermal radiation from an infrared heater (D). As implemented on the Alpha machine, the heater is a coiled nichrome wire that is mounted on a carriage. This allows the heating element to be passed over the surface of the powder bed. Steps A-D are repeated until the part is completed. In the end (Figure 8-2), a solid polymeric block remains that is totally sintered except for those areas wetted by the inhibitor liquid (E). The final part can be easily extracted from surrounding material (F).

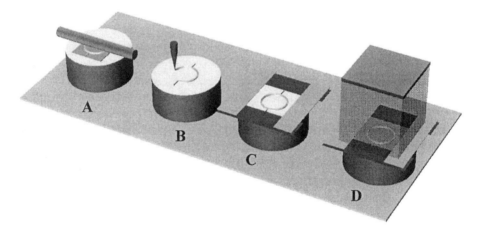

Figure 8-1. Stages of the SIS Process.

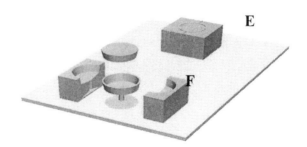

Figure 8-2. Extraction of the Fabricated Part.

Advantages of the SIS Process include:
- Low cost (the high power laser generator of the processes such as SLS is replaced with an inexpensive heat element).
- High speed (the entire layer is sintered at once in a few seconds), dimensional accuracy and surface quality (by use of high resolution inkjet printers in combination with fine powder particles).
- The possibility of having a large build tank which will allow for fabrication of large parts and small to medium lot size production of multiple parts in a single run (in contrast, SLS cannot build large parts because at the far boarders the laser needs to deflect beyond the extent feasible for accurate part edge fabrication.
- The possibility of fabricating multi-color parts, if various colors of the inhibitor agent are deposited.

An Alpha machine was built for initial research studies (Figure 8-3). The machine includes a source powder tank with a stepper motor driven piston which pushes the powder upward with the desired magnitude, a build tank that receives the powder and has a stepper motor driven piston which lowers progressively at the layer thickness increments, a roller that takes the powder from the source tank and spreads it over the build tank, an X-Y print head that moves the print nozzle over the desired layer profiles, and a drop-on-demand nozzle activated by an electro-magnetic solenoid valve that can operate at high frequency (up to 1000 Hertz) for printing the inhibitor liquid. The nozzle requires back pressure of at least 3 psi and can deliver droplets as small as 5 nanoliters. Note that this droplet size is far greater than typical droplet sizes generated by current inkjet printers. This specific nozzle was chosen because it could handle a wide range of liquid characteristics such as viscosity and surface tension. This flexibility was important, as numerous inhibitor liquid choices had to be evaluated. Some of the toxic inhibitors as well as some highly reactive liquids which deteriorated the nozzle had to be abandoned. Also a heat radiating element that can be positioned over the build tank to sinter the powder was used[9]. The SIS process includes the following major steps (see Figure 8-4):

0- The inhibitor tank is filled with the inhibitor liquid and the powder feed tank is filled with polymer powder, the inhibitor tank pressure and printer voltage are set, and the printer home is found. The choice of the printer nozzle is a miniature solenoid valve which requires back pressure. The voltage applied to the solenoid affects the response time and liquid pumping power.

1- The powder feed tank pushes the powder level upward in the amount of desired step(s). The build tank moves the powder level downward in the amount of desired step(s). The amount of lowering movement sets the layer thickness in the final part.

2- The cylindrical roller from the source tank spreads the powder over the build tank. The roller rotates clockwise and moves in the X direction.
3- An X-Y printer head moves the print nozzle over the desired layer profiles. The input file for the print pattern is the machine path file which is generated by the machine path generator system.
4- The heat source which is attached to the moving part of the machine (including the printhead and roller) moves in the X direction and its temperature and movement create a sintered layer with an unsintered pattern in the selected area of the powder bed. Steps 1-4 are repeated until the part is finished.
5- The model is removed from the sintered block and the post processing including cleaning, and for certain materials adding wax or adhesive to the fabricated part, completes the SIS process.

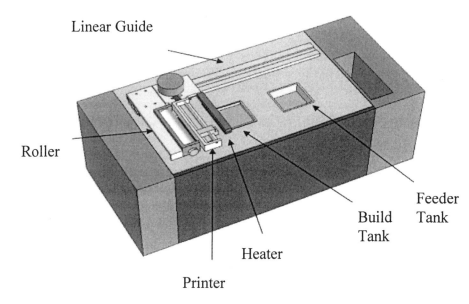

Figure 8-3. The Selective Inhibition Sintering Alpha Machine.

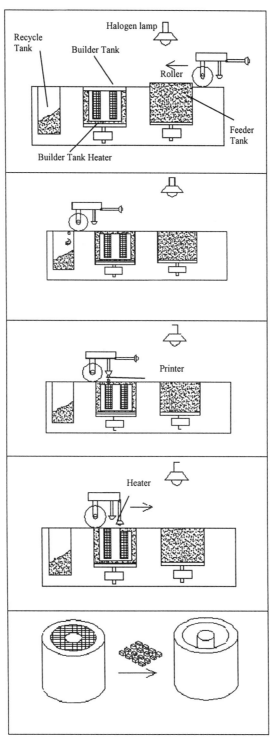

Figure 8. 4. Selective Inhibition Sintering Process Steps.

8.2. THE SIS PROCESS MATERIALS

Two types of materials are used in the SIS process: powder and inhibitor. The most favorable powder is the one that produces rigid parts with no deformation and warpage. Also, an ideal inhibitor is the one that, with minimum amount of material usage, can prevent the polymer powder from sintering. In addition to the technological requirements for the selected materials, these materials should also be non-toxic, easily accessible, and relatively inexpensive. Because of the newly developed idea of the polymer powder - inhibitor interaction, very few relevant literatures could be found. Therefore, most of the efforts for the material selection were based on the experiments and observations of the results for any potential materials. More than 20 different types of the polymer powders and more than 50 different industrial liquids and many more combinations of their mixtures were tested. Based on conducted experiments polystyrene polymer was the most appropriate polymer for making high quality parts (e.g., sharp edges, good attachment between layers, and low deformation). Also, use of salt solutions as the inhibitor has had the most successful results. Here the salt solution is printed as the inhibitor. After layer sintering and liquid evaporating, a thin layer of salt remains in the selected area. At the end of the SIS process, this wall can be dissolved and washed away by water (Figures 8-5 and 8-6). A small amount of alcohol is added to water to reduce the surface tension. This enables water penetration into the solid block. Because of the weak powdery bounding between polymer particles in the selected area, the final part can be easily extracted from surrounding material. Potassium iodide (KI) has high solubility (135 gr/ 100 gr water) and a high melting point (635° C). A saturated aqueous KI solution has been successfully used for many part fabrication experiments. Figure 8-7 illustrates steps for a sample part production using KI as the inhibitor[2].

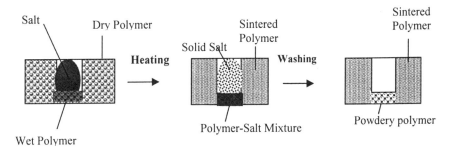

Figure 8-5. Water Evaporation in the Selected Area and Salt Wash Off to Extract the Part.

Figure 8-6. Salt Wash Off and Part Extraction.

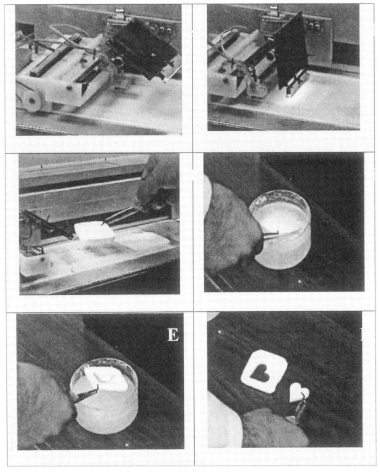

Figure 8-7. Steps for a sample part production using KI solution as the inhibitor: A) spreading the powder and printing the inhibitor, B) sintering (repeated steps A and B for 10 layers), C) removing the solid block from the build tank, D) Washing the block in a water-alcohol solution, E) dissolving KI salt into the water release part from surrounding material, and F) removing the part from the sintered block.

8.3. THE SIS PROCESS MACHINE PATH GENERATION

Because of the differences in the nature of Rapid Prototyping (RP) processes, there is no standard machine path code for them. Each RP process, based on its characteristics and requirements, uses the standard CAD file format to extract the required data for the process. Like other RP systems, SIS needs a specialized machine path generator to create an appropriate machine path file. Machine path (i.e. boundary path and hatch path) should produce the printing pattern that enables the SIS machine user to easily remove the fabricated part from the surrounding material. A new machine path algorithm, which generates appropriate boundary and hatch paths for the SIS process is developed. The new machine path generator provides the ability to process CAD models of any size and complexity, the ability for machine path verification before sending the file to the SIS machine, and the ability to fix the possible STL files disconnection errors. This system uses STL files with the ASCII format as input and works in two steps (Figure 8-8). The first step generates slices at each increment of Z by intersecting the XY plane with the facets within the part. The second step uses all individual unsorted intersection lines in each Z increment to form the contour. This step includes sorting the intersection lines, recognizing the closed loops and disconnections (STL file error), and generating the machine path (i.e. boundary path and hatch path) and simulation files.

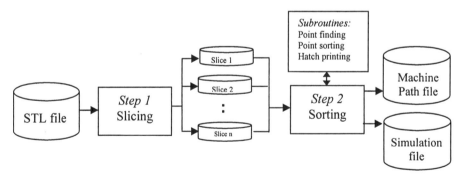

Figure 8-8. The Two Steps of Slicing and Machine Path Generation.

8.3.1 Step 1: Slicing Algorithm

In the first step the STL file is read as input. Slice files are then generated by executing the slicing algorithm. Only the intersection of those facets that intersect current Z=z are calculated and saved. In this step, one facet is read at a time from the STL file. Then the intersection lines of this facet with all

XY planes for $Z_{min} \leq z \leq Min\{Z_{max}, Max\{z_A, z_B, z_C\}\}$ are calculated. The intersection lines are stored in the specified file for the associated z increment. This results in one intersection line on each XY plane. By repeating this process for all facets, a set of slices is generated. This algorithm only saves the data of one facet in the computer memory, therefore only a small amount of computer memory is needed, and there is no practical limitation on the model size. In this step each slice is saved in a separate file on the disk. This guarantees that step 2 is run much faster as compared with the case where all slices are saved in a single file. The example shown in Figure 8-9 illustrates the slicing algorithm.

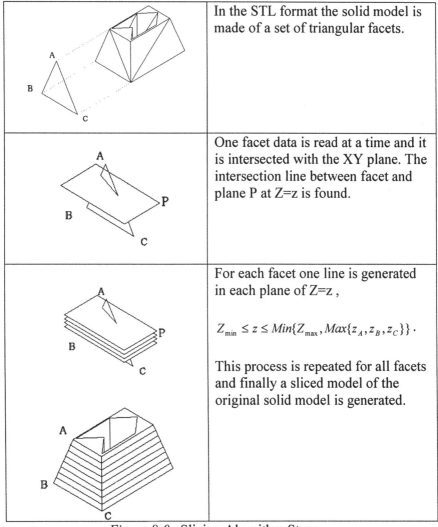

	In the STL format the solid model is made of a set of triangular facets.
	One facet data is read at a time and it is intersected with the XY plane. The intersection line between facet and plane P at Z=z is found.
	For each facet one line is generated in each plane of Z=z, $$Z_{min} \leq z \leq Min\{Z_{max}, Max\{z_A, z_B, z_C\}\}.$$ This process is repeated for all facets and finally a sliced model of the original solid model is generated.

Figure 8-9. Slicing Algorithm Steps.

8.3.2 Step 2: Machine Path Generation

After the completion of the slicing process, a set of vectors becomes available in each z increment. These vectors are not connected and are not in sequence.

In the machine path generation process, the software starts from one vector and tries to find the next connected vector to this vector. Then it does the same for the newly found vector until it reaches the start point of the first vector (in the closed loop cases) or finds a vector with no leading attachment (in faulty STL files containing disconnections). To sort the vectors, the algorithm reads one vector at a time from a slice file and writes it to another file. This file is either path file, when vector is connected to previous vector, or temp file, when the vector is not connected to the previous vector. Therefore, the sorting process does not need a large amount of memory to sort the data, and there is no limitation on the number of vectors in a slice and on input file size. In addition, unlike many other slicing algorithms that cannot handle disconnections caused by faulty facets[14], this algorithm can generate a machine path even with disconnection errors in the STL file. At disconnection instances the system sends a message to a log file and turns the printer off and starts from a new vector. In either case the printer is turned off and the system starts printing from another start point. Also for each selected vector the possibility of hatch intersection points are investigated.

At the end of the path generation process for one slice, the hatch intersection points are sorted and written into a machine path file. After the arrangement of all vectors in one slice (z increment), the process starts the arrangement of the vectors of the next slice. This process is continued until all vectors in all slices are sorted. Diagrams in Figures 8-10 and 8-11 represent the machine path generation algorithm. This algorithm includes boundary path and hatch path generation.

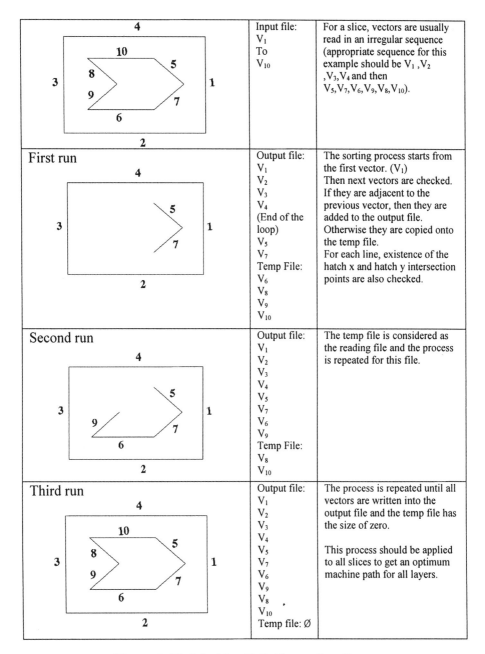

Figure 8-10. Machine Path Generation Steps.

To show the hatch path generation subroutines, basic steps are explained for the previous example.

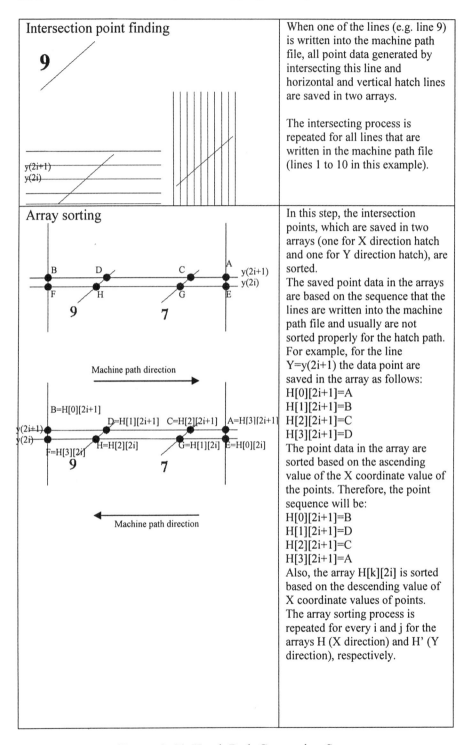

Figure 8-11. Hatch Path Generation Steps.

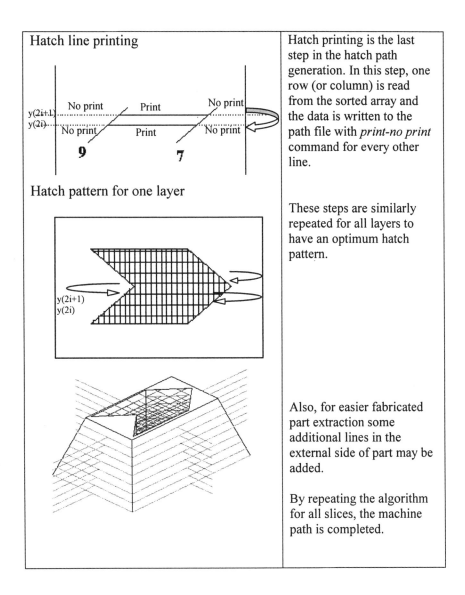

Figure 8-11. (Continued).

Figure 8-12 shows the algorithm implementation as presented by the path simulation module of the system. More details of algorithms can be found at Asiabanpour and Khoshnevis[4,5].

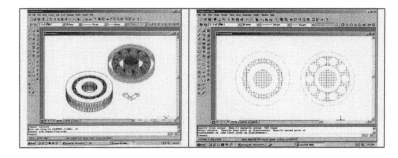

Figure 8-12. Visualization CAD model and the machine path (left) and two slices of the machine path.

8.4. THE SIS PROCESS OPTIMIZATION

Several properties (i.e., strength, surface quality, and dimensional accuracy) are important for the parts fabricated by the SIS process. Also, many different process operating conditions (factors) influence these properties. As shown in the Figure 8.13 about 30 factors are potentially effective on the SIS process in the form of IDEF0 diagram[12]. However, simultaneous controlling of all factors was impossible. Therefore, many of less effective factors were identified and eliminated experimentally or by methods such as cause-effect (fishbone) chart. *Layer thickness, Machine feed rate (roller), Printer feed rate, Printer frequency, Printer pressure, Heater feed rate, Heater temperature,* and *Build tank heater temperature* were eight factors that were selected for further control.

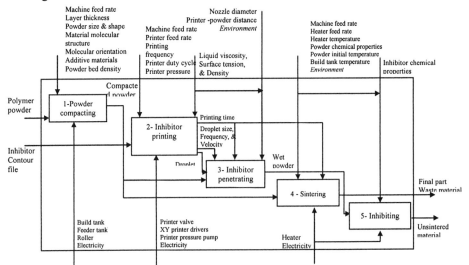

Figure 8-13. IDEF0 Hierarchy for the SIS Process.

After eliminating less effective factors and defining the range for each factor, statistical tools have been applied to improve the desirable properties of parts fabricated by the SIS process. Initial attempts to simultaneously optimize all responses for the SIS process produced several runs with missing data (immeasurable parts that were very brittle), and hence difficulties in applying statistical analysis tools. Consequently, it was decided that a region of operating conditions that provide acceptable part strength should be identified prior to investigation of conditions that improve surface quality and dimensional accuracy (Figure 8-14).

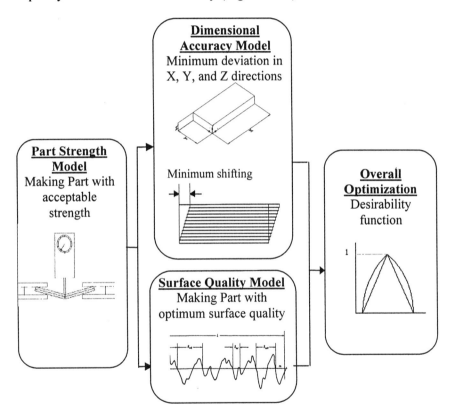

Figure 8-14. Roadmap for the SIS Process Properties Optimization.

Part properties improvement started with an investigation of part strength using Response Surface Methodology (RSM)[15]. For this purpose a sample block with dimensions $45 \times 20 \times 6$ mm was selected as the base part geometry for the experiments (Figure 8-15). Settings of the layer thickness factor produced specimen parts with differing numbers of layers; therefore, different machine path files were generated.

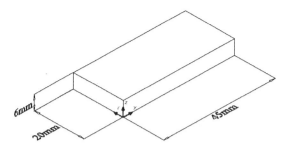

Figure 8-15. Base Part.

Then transverse rupture strength test, the standard basis for measuring the strength of the low ductility powder metallurgy materials (ASTM B528), was used as the response measure for the experiments (Figure 8-16).

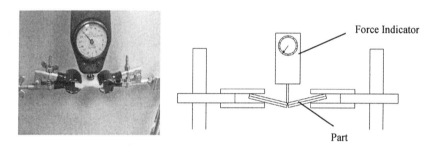

Figure 8-16. Part Breaking Mechanism.

Then, a mathematical model of part strength versus effective factors was developed and region of acceptable operating conditions for the part strength (part strength > 50 oz) was found (Equation 8.1).

$Strength = -441.3 + 597.4 \times LT - 6.11 \times HF + 1.095 \times HT - 0.862 \times BT + 2.442 \times HF^2$
$- 1.136 \times LT \times HT - 1.39 \times LT \times BT - 0.086 \times HF \times HT + 0.003 \times HT \times BT$
(Eq. 8.1)
Where:
LT: Layer Thickness
HF: Heater Feed rate
HT: Heater Temperature
BT: Build Tank Heater Temperature

Then, by using RSM, the impact of the factors on the final part surface quality and dimensional accuracy (for x,y, and z directions) was modeled. For this purpose first, quality criteria were identified. Minimum deviation between fabricated part dimension and CAD model dimension was selected as one of the part accuracy criteria. To measure the deviation, each axis (X, Y, and Z) was studied separately. For finding deviation of each axis, length (X), width (Y), and height (Z) values of the fabricated parts were measured using a caliper. Then, deviations from CAD model dimension were calculated as the error percentage. For example, deviation in the X axis was calculated through $E_X = \dfrac{X-45}{45} \times 100$, where X is the measured length of the fabricated part.

By observation of the fabricated parts from the previous experiments it was noticed that some of the parts had been made with noticeable shift in the X axis direction. To minimize such a defective property, layer shifting in the X axis direction was selected as another part accuracy response (named shift_X). Shift_X (total shift in X axis direction between first and last layer) was measured by a caliper in mm (Figure 8-17).

Figure 8-17. Shift_X.

Laser scanning systems are used to produce high precision surface quality evaluation. Mike Shellabear (1999) performed an extensive surface quality comparison between many commercial rapid prototyping systems. For surface quality measurement, he used a Zeiss Laser Scanning Microscope (LSM), which has a quoted accuracy of 0.2µm in X and Y directions and 0.6µm in Z direction.

In this research, a rating system was used as a low cost alternative to evaluate the part surface quality. For this purpose, ten sample parts with different surface qualities were selected and assigned a rating between 1 and 10 to indicate the worst to the best surface quality. Then, all fabricated parts were rated by comparison to the selected sample parts.

For precise rating, four boundary sides of each part are studied individually (Figure 8-18). Each side is rated based on three major properties: holes made of the accumulated salt (due to over printing), sharpness of the edges, and smoothness of the surfaces. At the end, an average value resulting from the 12 ratings (4 sides × 3 properties for each side) is assigned as the part surface quality rating (Table 8-1).

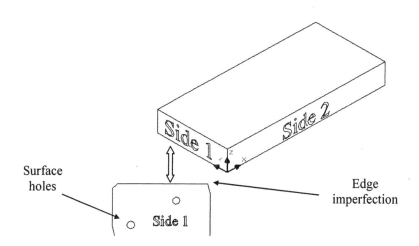

Figure 8-18. Part Breakdown for Surface Quality Rating.

Table 8-1. Part surface quality rating system

Part	Side 1			Side 2			Side 3			Side 4			Grade
No.	Edge	Smoothness	Holes	E	S	H	E	S	H	E	S	H	Sum/12
k	6	8	5	4	5	6	5	5	5	9	2	6	5.5
k+1	...	...	...	.	.	.					...	.	

The combined data set was used to develop mathematical models to show the responses behaviors (for E_x, E_y, E_z, Shift_X, and Surface quality) versus the factors (*Layer thickness, Machine feed rate (roller), Printer feed rate, Printer frequency, Printer pressure, Heater feed rate, Heater temperature,* and *Build tank heater temperature*) and significant terms of the second order polynomial models were identified by stepwise regression.

By analyzing the developed mathematical models for E_x, E_y, E_z, Shift_X, and Surface quality, it is known that there are some tradeoffs among these responses. This means that increasing the value of one factor might produce a more favorable result for one response but would produce a less favorable result for another response. To balance this tradeoff, a multi-response optimization technique is needed. Desirability function[11] is used to balance the tradeoff between responses. General form of the component desirability function for each response can be illustrated as the one shown in Figure 8-19 and Equation 8.2.

(Eq.8.2)
$$d = \begin{cases} 0 & , \quad \hat{y} < a \\ \left(\dfrac{\hat{y}-a}{b-a}\right)^s & , \quad a \leq \hat{y} \leq b \\ \left(\dfrac{\hat{y}-c}{b-c}\right)^t & , \quad b \leq \hat{y} \leq c \\ 0 & , \quad c < \hat{y} \end{cases}$$

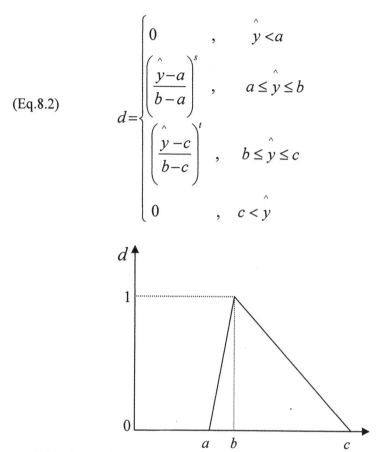

Figure 8.19. General Form of the Component Desirability Function.

The response value labeled b in Figure 8-19 represents the target quality. This response value corresponds to the maximum desirability ($d=1$). In theory, the response values labeled a and c represent the limits of acceptability. However, in practice, the values for a and c should be selected to produce a minimal number of infeasible locations in the factor space. Infeasible locations are defined by $d=0$. (for simplicity, we select: $s=1$ and $t=1$ (linear equation)). After developing a desirability function model, process operating conditions for maximum desirability were identified.

Finally, the desirability model was validated. To do so, some additional sample parts were fabricated. All five (E_x, E_y, E_z, Shift_X, and Surface quality) responses were measured for each part and average results were calculated. For each response, the individual response models were used to calculate 95% prediction intervals for the average of values. All actual averages fall within their respective prediction interval. These results indicate that the models are valid. More details of the SIS process optimization can be found at Asiabanpour, Palmer, and Khoshnevis[6,7,8].

8.5. PHYSICAL PART FABRICATION

Several 2.5 and 3D parts have been successfully fabricated by the SIS process using the machine and hatch path generation system. Figures 8.20 and 8.21 illustrate some of the fabricated parts.

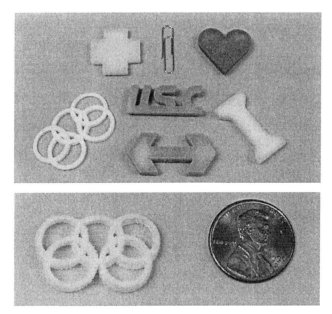

Figure 8- 20. 2.5D Parts Fabricated by the SIS Process.

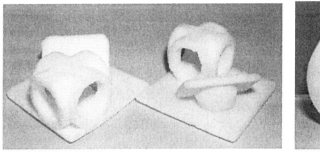

Figure 8- 21. 3D Parts Fabricated by the SIS Machine.

8.6. POWDER WASTE REDUCTION

The heaters used in the current SIS process apply heat to a fixed area of the build tank surface. The end result of applying heat to the entire fixed area is that large amount of support polymer powder is being wasted. Figure 8.22 shows some parts made from the SIS Alpha Machine. Contrasting colors were used to indicate polymer powder waste. A great deal of polymer powder can be saved by selectively masking areas, which then would make the SIS process far less expensive.

Figure 8-22. Current SIS Process Limitations: Waste Material.

Various means should be devised and studied for powder waste minimization. A combination of hardware and software approaches can significantly reduce powder waste in SIS. A systematic approach that has been applied to design a new modular, reconfigurable heater allows for the adjustment of the masked area in order to selectively heat the required areas. For example, in Figure 8-23 to make a part (figure 8-23(a)), entire square area (figure 8-23(b)) is sintered while only small portion of it (circle area) is useful. Figure 8-23(c) illustrates mask system and dark area in figure (8-23(d)) shows the waste area.

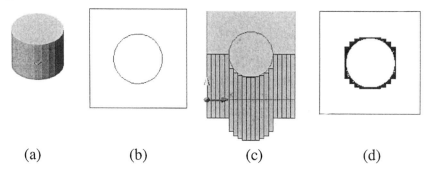

Figure 8-23. Moveable Fingers Mask System.

The conducted systematic design approach includes simulation studies, alternatives generation, concept selection, design for manufacturing and assembly, manufacturing and assembly, and test. The study revealed that finger based masking system design and prototypes are very promising from waste saving standpoint. In this system which depends on the profile of each layer of the part, small heat resistant fingers move to such a position that the unwanted sintered powder (waste) is kept to a minimum. Figure 8-24 shows more details of the mechanical structure of this system. In this system two perpendicular rails give the flexibility to adjust the configurations for variety of profiles. MDF (medium density fiberboard) was selected as the material on which to mount the aluminum base because of its non-flammability property. In this system one motor in Y axis pushes a finger to the assigned position. Then, another motor moves this motor to a position such that the next finger can be moved. When a finger is required to be pulled back, a magnet system is activated for that finger which is made out of an electromagnetic base and MDF mask[3].

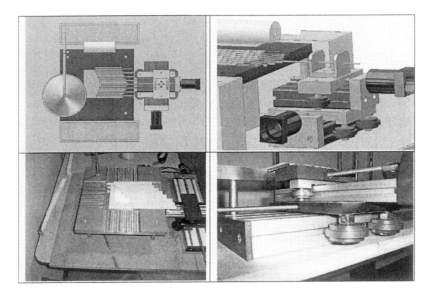

Figure 8-24. The New Mask System Design (top) and Prototype (bottom) for the SIS Rapid Prototyping Process.

8.7. VISION FOR FUTURE RESEARCH

The multidisciplinary nature of the SIS process development opens a wide variety of areas for future work.

    I. Hardware issues

        Several changes in the original SIS machine structure could result in process improvement. Applying inkjet printer instead of a single nozzle printer makes the system much faster and more precise. Also, a larger build tank provides the ability to fabricate larger parts. In addition, improvement in heater system can cause more uniform layers and consequently result in more accurate parts.

II. CAD/CAM issues

Current machine path development is specialized for the single nozzle printer. In case of applying inkjet printer, a CAD/CAM system is required to produce raster image, which include boundary paths and hatch patterns. Raster images can also provide data for color printing to make multi-color parts.

III. Material issues

This research has focused on a limited number of polymer powders. Based on required part properties, many other polymers should also be applicable. The SIS process is also expected to be applicable to metallic powders. We are currently investigating the use of copper, steel, and aluminum powders. These powders may require the use of inhibitors that capitalize on inhibition phenomena other than particle separation at the microscopic level. For example, inhibitors that produce chemical reactions with the powder, or those that produce particle separation at the macroscopic level may be considered. Finding appropriate inhibitors to allow the use of powders that produce a variety of part properties is a broad area of research.

IV. Powder compaction issues

The fabrication of uniformly dense parts can be facilitated by powder compaction prior to sintering. This is especially true for metallic powders. However, the presence of the inhibitor in the powder layer complicates the compaction process, specifically in terms of the dimensional stability of the layer. The extension of existing powder compaction models[1,10,13] to the case of compaction for powder layers containing 'interior walls' of inhibitor material remains an open area for future work.

V. Process optimization issue

A systematic approach was explained in this article to optimize the SIS process settings. However, all related factors were not considered in the statistically designed experiments. Consideration of other factors (e.g., material related factors) as well as application of other optimization methods (e.g., Taguchi) may lead to more realistic models.

References

1. ABAQUS/Standard User's Manual, Version 5.7, Volume 1-3, Hibbitt, Karlsson, & Sorensen, Inc., Providence, RI, (1997).

2. B. Asiabanpour,.An experimental study of factors affecting the Selective Inhibition of Sintering process, Ph.D. Thesis, University of Southern California, Los Angeles, CA (2003).
3. B. Asiabanpour, F. Wasik, R. Cano, V. Jayapal, L. VanWagner, and T. McCormick, New waste-saving heater design for the SIS Rapid Prototyping Process", IERC Conference, Atlanta, GA (2005).
4. B. Asiabanpour, and B. Khoshnevis, A new memory efficient tool path generation method for applying very large STL files in vector-by-vector rapid prototyping processes, Computers and Industrial Engineering Conference, San Francisco, CA (2003).
5. B. Asiabanpour, and B. Khoshnevis, Machine path generation for the SIS process, *Journal of Robotics and Computer Integrated Manufacturing*, **20**(3):167-264 (2004).
6. B. Asiabanpour, K. Palmer, and B. Khoshnevis, Performance factors in the Selective Inhibition of Sintering process, IIE Conference, Portland, OR (2003).
7. B. Asiabanpour, K. Palmer, and B. Khoshnevis, An experimental study of surface quality and dimensional accuracy for selective inhibition of sintering, *Rapid Prototyping Journal*, **10**(3):181-192 (2004).
8. B. Asiabanpour, B. Khoshnevis, K. Palmer, and M. Mojdeh, Advancements in the SIS process, 14th International Symposium on Solid Freeform Fabrication, Austin, TX (2003).
9. B. Khoshnevis, B. Asiabanpour, M. Mojdeh, and K. Palmer, SIS – A New SFF method based on powder sintering, *Rapid Prototyping Journal*, **1**(1):30-36 (2003).
10. D. V. Tran, R.W. Lewis, D.T. Gethin and A.K. Ariffin, Numerical modeling of powder compaction processes: displacement based finite element method, Powder Metallurgy, **36**, 257-266(1993).
11. G. Derringer, and R. Suich, Simultaneous optimization of several response variables, *Journal of Quality Technology*, **12**(4):214-219 (1980).
12. Integrated DEFinition method website, http://www.idef.com.
13. J.L. Chenot, F. Bay, and L. Fourment, Finite element simulation of metal powder forming. *International Journal of Numerical Methods Engineering*, **30**,1649-1674 (1990).
14. K.F. Leong, C.K. Chan, and Ng, Y. M., A study of stereolithography file errors and repair, *The International Journal of Advanced Manufacturing Technology*, 407-414 (1996).
15. R. Myers, D. Montgomery, *Response Surface Methodology*, John Wiley & Sons, Inc (2002).

Chapter 9

CONTOUR CRAFTING
A Mega Scale Fabrication Technology

Behrokh Khoshnevis[1] and Dooil Hwang[2]
[1] Professor, Department of Industrial and Systems Engineering, University of Southern California, USA; [2] Research Associate, Information Sciences Institute, University of Southern California, US

Abstract:

Contour Crafting is a mega scale fabrication technology based on Layered Manufacturing process (LM). This fabrication technique is capable of utilizing various types of materials to produce parts with high surface quality at high fabrication speed. The process extends its fabrication capabilities to construction of houses and civil structures.

Key words:

Contour Crafting; Layered Manufacturing; large scale fabrication, mega scale fabrication, extrusion, troweling, formative process, thick layer fabrication, thermoplastics extrusion, ceramics extrusion, concrete extrusion, construction automation; house construction; Lunar construction, extraterrestrial construction,

9.1 INTRODUCTION

An innovative rapid prototyping process, called Contour Crafting (CC), has been developed at the University of Southern California (USC). Because of its unique capability in using thick layers with smooth surface quality, the CC process is suitable for rapid fabrication of large-scale complex shaped objects with smooth surface finish[1]. The CC process is based on an extrusion and filling process shown in Figure 9-1.

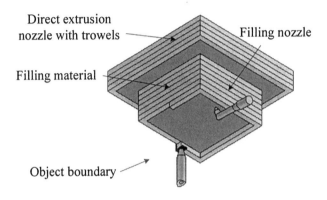

Figure 9-1. Schematic of CC extrusion and filling process

The extrusion process forms the smooth object surface by constraining the extruded flow in the vertical and horizontal directions by the use of trowels. A schematic view of extrusion using two trowels is shown in Figure 9-2.

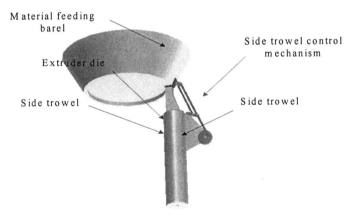

Figure 9-2. Schematic of trowels and extrusion assembly

The orientation of the side-trowel is dynamically changed for better surface fit for each decomposed layer. The side-trowel allows thicker material deposition while maintaining high surface finish. Thicker material deposition cuts down manufacturing time, which is essential for building large-scale parts using the material additive process. Maximum deposition layer thickness is limited by the trowel height.

As the extrusion nozzle moves according to the predefined material deposition path of each layer, the rims (smooth outer and top surface of outside edges) are first created. The toweled outer surface of each layer determines the surface finish quality of the object. The smooth top surface of each layer is also important for building a strong bond with the next layer above. Once the boundaries of each layer are created, the filling process begins and material is poured or injected to fill the internal volume.

9.1.1 CC Process error analysis

Due to no having control of the outer surface of layers, parts fabricated by current LM techniques generally have poor surface finish, especially for slant surfaces which demonstrate a stair casing effect. Another reason for poor surface finish is process error. Examples of stacked layers of parts are shown in Figure 9-3.

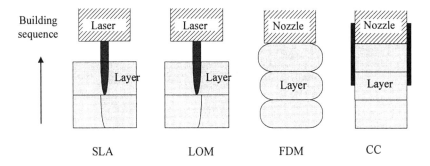

Figure 9-3. Cross-section comparisons of layered boundaries

One of the great advantages of the CC process is its superior surface forming capability. A mathematical model for process error was introduced to define the fabricating orientation based on the minimum process error[2].

The CC process has zero process error in fabricating 2.5D geometries and some 3D geometries. The layered process error (e_i) is obtained from accumulation of deficient filling error and over filling error between the layers (v_i) and (v'_i),

$$e_i = (v_i - v'_i) \cup (v'_i - v_i) \quad (1)$$

From equation (1), the total layered process error for an n-layered object can be express as:

$$E = e_1 \cup e_2 \cup \ldots \cup e_n. \quad (2)$$

Local layer process error is shown in Figure 9-4. A curvilinear boundary AC can be approximated as a straight surface for thin layer and then the area of the triangle ABC represents a differential layered process error (de_i).

$$de_i = S_{\triangle ABC} = \frac{h_i^2 \cos\alpha}{2\sin\alpha} \quad (3)$$

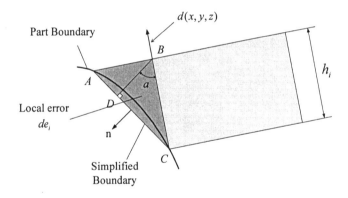

Figure 9-4. Geometric description of local layered process error

According to the definition of gradient function[3], its modules $|\mathbf{d}(x,y,z)|$ are equal to the reciprocal of layer thickness:

$$|\mathbf{d}(x,y,z)| = \frac{1}{h_i} \quad (4)$$

According to Figure 9-4 the following can be obtained:

$$\cos\alpha = \frac{|\mathbf{d}(x,y,z)\cdot\mathbf{n}|}{|\mathbf{d}(x,y,z)|\cdot|\mathbf{n}|} \qquad (5)$$

$$\sin\alpha = \frac{|\mathbf{d}(x,y,z)\times\mathbf{n}|}{|\mathbf{d}(x,y,z)|\cdot|\mathbf{n}|}$$

Substitute equation (4) and (5) into (3):

$$de_i = \frac{h_i^2\cdot|\mathbf{d}(x,y,z)\cdot\mathbf{n}|}{2|\mathbf{d}(x,y,z)\times\mathbf{n}|} \qquad (6)$$

Integrating equation (6) along the entire contour of the $i\,th$ layer, the layered process error can be obtained from:

$$e_i = \oint de_i \cdot dl = \oint \frac{h_i^2\cdot|\mathbf{d}(x,y,z)\cdot\mathbf{n}|}{2|\mathbf{d}(x,y,z)\times\mathbf{n}|}\cdot dl \qquad (7)$$

Therefore, the total layered process error can be expressed by:

$$E = \sum_i e_i = \sum_i \oint \frac{h_i^2\cdot|\mathbf{d}(x,y,z)\cdot\mathbf{n}|}{2|\mathbf{d}(x,y,z)\times\mathbf{n}|}\cdot dl \qquad (8)$$

From the total layered process error, if the gradient of fabrication orientation $\mathbf{d}(x,y,z)$ is perpendicular to normal vector $\mathbf{n}$, the total layered process $\mathbf{E}$ will be mathematically equal to zero because $|\mathbf{d}(x,y,z)\cdot\mathbf{n}|$ will be equal to zero.

In the CC process, the normal vector $\mathbf{n}$ is perpendicular to the gradient of fabrication orientation $\mathbf{d}(x,y,z)$ for fabricating 2.5D geometries and some 3D geometry such as the cone shape. Some 2.5D geometries and a 3D geometry are shown in Figure 9-5.

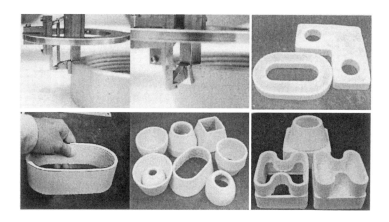

Figure 9-5. Contour Crafting nozzle and some of 2.5D geometries and 3D parts

9.1.2 CC applications

Contour Crafting has been the focus of intensive research at the USC rapid prototyping laboratory. The laboratory has been improving CC's superior surface forming capability, speed, part size and nozzle designs. Various materials have been tested and evaluated to date. Rapid advances in this research will be a critical if CC is to be considered as a viable option for fabrication of large scale products.

9.1.2.1 Ceramic part fabrications

Clay is a ceramic material which is abundant in nature and hence it has been used since the dawn of civilization for creating various objects with or without thermal treatment. Clay has been used in uncured form as clay bricks in construction. Sintered clay has been used for pottery and tiles for many centuries. More recently, ceramic material are finding advanced applications in manufacturing. Ceramics processing usually starts with ceramic paste the property of which is influenced by its mineral and structural composition, and by the amount of water that it contains.

Mechanical properties of ceramic paste primarily depend on the water content. With sufficient water the paste softens and forms slurry that behaves as a viscous and formable liquid. When the water content is gradually reduced by drying, the ceramic paste loses its plastic state property and holds together resisting deformations. Ceramic paste shrinks and its stiffness increases until it becomes brittle with further loss of water and enters the semisolid state. By drying continuously, the clay reaches a constant minimum volume at its solid state.

9.1.2.2 CC machine structure for ceramics processing

An assembled CC machine for ceramics processing is shown in Figure 9-6. The machine mainly consists of an extrusion unit and the trowel control mechanism. The extrusion unit carries uncured ceramic paste into material carrying tank and a linear ball screw driven piston pushes the paste through a CC extrusion nozzle. With controlled rotational speed of feeding motor, stabilized extrusion flow can be achieved. When complex shape of geometry is being fabricated, the system controls the angle and orientation of the side trowel to conform to outside surface geometry each cross sectional layers.

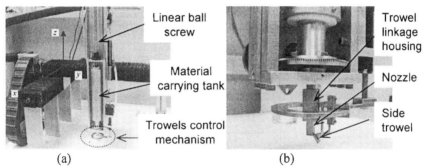

Figure 9-6. CC system configurations for ceramic part fabrications; (a) CC extruder sits on an x-y-z gantry robot, (b) details of CC nozzle

9.1.2.3 Preparing ceramic paste

The ceramic material (clay) used in our experiments was procured from America Ware Company in Los Angeles. The clay contained: Pioneer Talc 2882, Taylor Ball clay, Barium Carbonate, Soda Ash, Sodium Silicate, and 35% water by mass. The clay parts were fabricated at room temperature and then bisque-fired in a kiln at $1063^0C \sim 1066^0C$ for 10 hours. For glazing, a second firing at 1003^0C was carried out for 8~9 hours. Forced drying by heating and the wax-coating procedure were not necessary due to the higher structural wet strength of the clay material. Large parts could be made with the assurance that the clay would not sag or collapse inward.

A way of loading clay into material carrying tank was devised in order to avoid entrapping air inside the tank. Entrapped air typically causes some defect modes at fabricated parts as forms of voids, excessive dry shrinkage, weak structural integrity, etc. A filling method was devised using a funnel shape apparatus that enabled continuous insertion of clay into the tank and pre-extrusion processing of the material also seemed to significantly reduced the void and defect surface formations.

9.1.2.4 Prefabricated ceramic parts

Several pre-designed experiments were conducted to study the feasibility of CC construction process by fabricating various geometries, creating hollow depositions by application of mandrills inside the CC nozzle, and trying various concurrent processes including imbedding of reinforcement material such as steel coils. Figure 9-7 shows the some results of these trials. The experimentation with ceramic materials has demonstrated the possibility of applying the CC process to the field of construction.

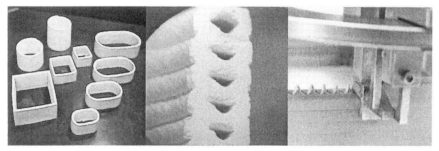

Figure 9-7. Demonstrations of CC constructability; (left) simple geometries, (middle) cross section of the fabricated part revealing hollow depositions, and (right) metal coil imbedding

9.1.2.5 Fabrication of pre-functional piezoelectric lead zirconate titanate (PZT) ceramic components and thermo-plastic parts

The goal of this project is to fabricate pre-functional piezoelectric lead zirconate titanate (PZT) ceramic components by adapting the Contour Crafting process. Piezoelectric materials are widely used in transducer, sensor, and actuator applications. A 1/16-inch diameter ceramic loaded polymer filament with 52 to 60 vol. PZT is used as feed material. Fused Decomposition of Ceramic (FDC), developed at Rutgers University, is used for fabricating ceramic prototypes by hot extrusion[4].

PZT-loaded polymer is extruded to fabricate green ceramic structures by a heated extruder head capable of moving in the X-Y direction. Then, the structures is subjected to a binder burn out cycle and sintered. The FDC system deposits four different materials in a single deposition step. This allows easy fabrication of composite components that have the property advantages of multi-materials.

To fabricate some of PZT ceramic actuator components shown in Figure 9-8, the research focused on the adaptation of CC process for fabrication of a thin wall structure. The CC process originally utilized one-sided trowel to control outer surface finish of rims; however, construction of thin walled structures also required a smooth inside surface finish.

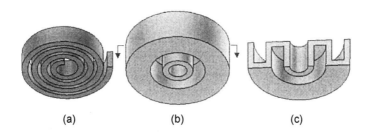

Figure 9-8. Advanced ceramic structures. (a) Spiral actuator, (b) Telescoping actuator, (3) Section of telescoping actuator

To fabricate advanced thin wall structures, a new CC system was designed assembled in the Computer Aided Manufacturing Lab at USC to meet the challenges. The CAD design of the new CC system and the final actual system are shown in Figure 9-9.

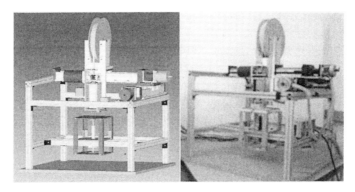

Figure 9-9. An adapted CC system for fabricating pre-functional ceramic and thermoplastic parts

9.1.2.6 Design considerations

For the new CC system, the filament material acts as billet to provide the necessary pressure to extrude the filament out of the nozzle. One of the design considerations for the new system was buckling of the filament due to a high viscosity and to a low stiffness. To eliminate this problem, the feed system was designed to allow length changes (L). It would have been too difficult to obtain this parameter from mathematical analysis. Euler's buckling criteria for elastic columns is used in the design cycle[5].

$$\sigma_{cr} = \frac{\pi^2 E}{4(L/R)^2} \tag{9}$$

where:

σ_{cr} - critical buckling stress
E - compressive stiffness of filament
L - length of the columns
R - radius of the filament

In the equation, the length change has strong effect on the critical buckling stress and the design reflects the variable length between the feeding rollers and the heating barrel. The picture of the feed system assembly is shown in Figure 9-10.

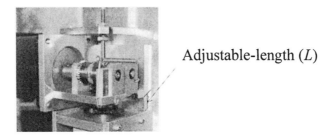

Figure 9-10. Material feeding system with adjustable length (*L*) between rollers and heating barrel

Another consideration was to design the working platform where the physical part is build up. When the fused material is extruded from the computer controlled nozzle, the heated extrudate needs to adhere to a base platform; otherwise, it will stay and follow the path of the trowels. The functional requirement of the base platform is to hold the first layer of a rim tightly and to prohibit thermal contraction. Thermal contraction causes extrudate separation from the base platform.

To best meet the functional requirements of a base platform, a thermoplastic plate was selected and thermoelectric heat generators installed under the surface. The heaters supply enough heat for keeping the extrudate in a fused state. While the first layer is in the fused state, the remaining layers are deposited successively on the top of previous rims until the part is completed. The completed part is easily removed from the base platform due to thermal contraction which occurs after the heaters are turned off. The base platform schematic is shown in Figure 9-11.

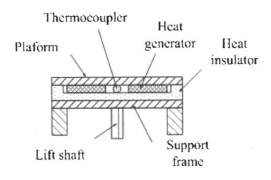

Figure 9-11. Schematic of CC working platform to prevent thermal contraction

9.1.2.7 Experiments

Ceramic loaded polymer filament was provided from Rutgers University. This length allowed only about three fabrication attempts of the parts. To save this limited resource, a thermoplastic (P-400) provided form Stratasys inc., was used to optimize process parameters and to understand the flow behavior.

The initial experiments conducted with fabrication of 2.5D geometries such as cylinders, spirals, and clovers were for developing a special die shape to control extrudate swelling and thermal contraction. The distorted square orifice shape was progressively designed with trial experiments. Schematic of die shape design is shown in Figure 9-12. Finally, a highly customized die shape was generated to deposit the high visco-elastic material while compensating geometrical accuracy and controlling the flow direction. The actual die shape used for fabrication of parts is shown in Figure 9-13. As the picture shows, the die has a sharp edged blade to generate a groove on the top of extrudate. The space defined by the groove boundary will be filled with the material as the thermal contraction progress.

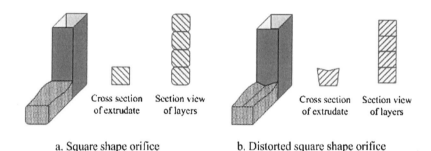

Figure 9-12. Schematic of die shape designs and sectional view of layers

Figure 9-13. Showing customized nozzle shape and its actual fabrication

Rapid Prototyping: Theory and Practice 233

With the distorted square die shape, relatively accurate 2.5D geometries and two advanced structures are fabricated with minimal swelling and thermal contraction of extrudate. The fabricated geometries with 2.5mm width are shown in Figure 9-14.

To fabricate the advanced spiral shape ceramic structure, one continuous tool path is used to avoid "stop and start" problems common in extrusion based RP systems. There can be poor surface finish after a start and stop event, due to the complexity of controlling extrudate flow. When the extrusion nozzle reaches one end of the first layer, the base platform moves down the height of the next layer thickness and the second layer is created. Fabricated spiral ceramic parts are shown in Figure 9-15.

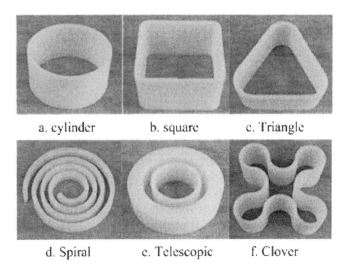

a. cylinder   b. square   c. Triangle

d. Spiral   e. Telescopic   f. Clover

Figure 9-14. Fabricated 2.5D geometries and two advanced structure shapes (spiral, telescopic)

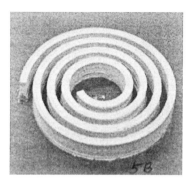

Figure 9-15. Pre-functional advance ceramic (Spiral PZT actuator)

9.1.3 Construction Automation

Since the CC process was developed in 1996, applications have been sought in areas such as automotive, aerospace, construction industries, etc. The common requirement of these industries is a need to fabricate large-scale components using different materials. Most commercialized RP systems cannot provide reasonable solutions for these issues, however the CC process has great potential. The CC process has the capability to fabricate exceptionally large components, quickly, and with high surface finish. The CC process has already been demonstrated for automated construction. For instance, it is possible to extrude materials with hollow cavities, reinforcements and impregnation, and sandwich structures[6,,7,8,9]. These geometric features are common in the construction industry. The experimentally proved CC technology is the most suitable process for fully automating manual construction works. Figure 9-16 shows one proposed application of the CC process for automated construction. As shown, the adobe housing structure is a good candidate since these structures minimize the use of support elements.

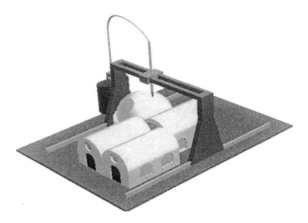

Figure 9-16. Schematic of the automatic construction of a residential building [source:contourcrafting.org]

9.1.3.1 Challenges faced by construction industry today

The construction industry presently faces challenging problems in the areas of productivity, quality, safety, and skilled labor shortages[10]. In order to address these issues, various types of automation and robotics technologies have been proposed and implemented. These attempts however have not returned significant improvements and little has been contributed to solving the problem. A major obstacle has been that the construction industry is steeped in high skill, labor intensive conventional processes, not conducive to adaptation of automation technologies.

CC is a specialized extension of the Rapid Prototyping (RP) technology for construction applications. In CC, full scale final structures (not prototypes) are manufactured at high speed with a smooth controlled surface finish. The primary merit of RP is the capability of fabricating very complex structures at lower costs and cycle times than conventional methods. This advantage has opened up many potential applications in a variety of fields. CC technology uniquely adapts RP capabilities and extends them to the field of large scale construction.

Contour Crafting represents a revolutionary paradigm in automation for the construction industry. CC replaces conventional "beam and post" methods with a layer-by-layer approach. This new way of thinking gives automation a much better chance to penetrate and succeed in the construction field[11].

9.1.3.2 The current state of automation in construction sites

There are two categories of automation considered by construction companies. The first is using single task robots that can replace simple labor activities at the construction sites. Single task robots can be classified by four different types- concrete floor finishing, spray painting, tile inspection, and material handing. Approximately 89 single task construction robots including welding have been prototyped and deployed in construction sites in Japan[12]. Figure 9-17 shows a deployed single task robot. This robot (named Surf Robo) finishes the concrete floor automatically. The Surf Robo is controlled wirelessly, fully automatic, and can finish concrete floors up to 300^2 m/ hour with double finishing.

Figure 9-17. A single task robot, Surf Robo, Copyright Takenaka Corp.

The second category consists of fully automated systems that can construct high raised steel buildings or steel reinforced concrete buildings using prefabricated components. Figure 9-18 shows the world's first automated construction system for building a precisely defined concrete structure. The big canopy shown in the figure is supported by four independent masts and has overhead cranes which deliver components at the control of a simple joystick. All tasks are scheduled and controlled by a centralized information control system shown in Figure 9.18a.

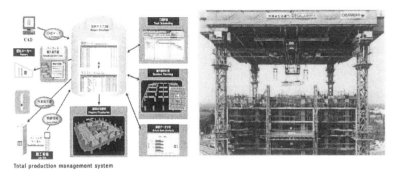

a. A centralized production management system   b. Big Canopy system

Figure 9-18. The first automated construction system applied steel concrete structure, Copyright Obayashi

The introduction of robotics directly at construction sites has contributed to productivity, safety, and quality improvements. Yet, the contribution of robotics at current levels is not revolutionary and current automation approaches are still geared toward conventional processes. Automating conventional processes is invariably expensive; so, the cost savings is minimal. Fast changing construction processes and project complexities create complicated requirements and exceptional challenges for automation technology to adapt.

9.1.3.3   Barriers for construction automation

There are many hurdles that automation technology must overcome. The following obstacles present challenges to efficient implementation of automated construction technology[13]:

- **Unavailability of automated fabrication technologies for large scale products**: There is no existing technology capable of fabricating large structures. New automation technology must be integrated with current large scale fabrication capabilities.
- **Conventional design approaches not suitable for automation:** Conventional construction processes and sequences are not suitable for automated systems. Any product manufactured using an automated system, must have a design and assembly sequence that is planned in advance to fit the automated system capabilities.)

- **Small production quantity compared to other industries**: Automation is a solution when identical or similar products are mass produced. Implementation of an automation system can be justified only by comparing the initial nonrecurring investment to the product's long term recurring profits.
- **Limitation of material choices:** There is a huge number of construction materials and the choice is generally determined based on economy. There is no one unique standardized material, therefore, automated systems must be nimble and have the capability of adapting to different materials.
- **High equipment cost and maintenance fee**: Automation require a huge capital investment and a specially trained operator. When a new project is started, the automation system needs to be reprogrammed to meet the new requirements. This requires a great amount of man power. Furthermore, automation systems, which operate in dust and dirt heavy construction sites, need frequent cleaning and maintenance.
- **Management issues**: All construction activities must be monitored in real time to meet deadlines; time is money in construction. A management system needs to be in place that monitors the running inventory of materials, assesses schedules, weather impacts, and progress on each task.

Previous attempts at developing automated construction systems could not adequately address these problems because of inherent barriers in the construction process. Modern automation technologies and engineering know-how may not be sufficient to overcome these barriers cost effectively. The construction industry should nevertheless seek alternate approaches to existing fabrication and assembly processes. The new approach should consider adaptability to available automated technologies and process

9.1.3.4 The needs for an innovative construction process

Previous attempts at automation by the construction industry have resulted in insignificant returns in productivity, safety, quality. The current state of automation and robotics technology, however, is not sufficient to economically replace skilled labor. The construction industry needs to think "out of the box" and seek alternatives to existing fabrication and assembly process. The new approach must consider adaptability to existing automated construction technologies and processes.

In recent years the Rapid Prototyping (RP) process has been implemented in a variety of applications and disciplines such as architecture, automobile design, aerospace and medical industries. Rapid prototyping (RP) is a material additive manufacturing process by which objects are created by computer controlled, layer by layer sequential material deposition. By using Computer Aided Design (CAD) packages, a 3D solid model of a part is decomposed into numerous cross-sectional layers. The cross-sectional geometries are mapped to material layer deposition paths, and fed to the RP machine where the physical 3D solid model is created in a layer by layer fashion. RP processes have significantly reduced the high cost and cycle-time of new product design and manufacturing. One goal of using the RP process is to verify the form and fit of new design concepts early in the product development stage, prior to final production[14].

The main merit of the RP is the capability of fabricating complex structures. However, RP systems today are not suitable for fabrication of larger scale parts. The CC technology adapts RP capabilities and extends them to the field of large scale fabrication. As in other RP methods, in Contour Crafting (CC) material is added layer by layer according to a computer controlled sequence.

9.1.3.5 CC concrete formwork design

A concrete wall form typically consists of sheathing, studs, wales, ties, and bracing as shown in Figure 9.19. A closer section of a traditional wall form is shown in Figure 9.20a. The fresh concrete is confined to the sheathing and places a lateral pressure on the sheathing until the concrete is cured.

Due to high lateral pressures, the sheathing is supported by properly spaced studs and wales to prevent any displacements. The tie rod is secured with nuts and washers to retain the wales and prevent bulging. The spreaders also help to maintain uniform wall thickness. After the concrete sets, the exposed part of the tie rod can be removed or kept in place as an integral part of the wall. The spreaders can also be removed or kept in place and mortar can be used to backfill the voids in the concrete if necessary.

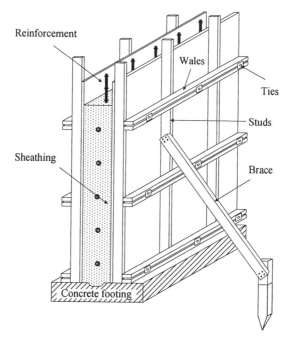

Figure 9-19. Basic components of a typical wall formwork

For the CC formwork design (Figure 9-20b) side walls (forms) are constructed using mortar and secured with U-shaped tie rods. This new formwork is simpler than the traditional design since it uses only two components: sheathing and tie. Sheathing is created in position by adding mortar continuously according to pre-defined material deposition sequence; the ties are inserted at the sheathing locations. The sheathing physical properties could be inferior to that constructed according to the traditional formwork system. The advantage however is that the new formwork can be constructed without using separate formwork materials. If the pour rate is less than 5 inch per hour (13 cm/hour), the lateral pressure is minimal and ridge formwork materials are not necessary. Sheathing and ties are kept in place as an integral part of the concrete wall after hardening.

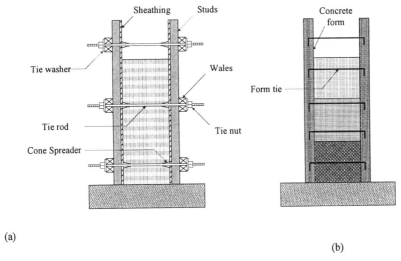

Figure 9-20. Closer sections of wall form; (a) traditional wall form, (b) CC wall form

9.1.3.6 Physical properties of CC formwork material

For conventional cast in place concrete construction, the formwork is manufactured using wood, plywood, steel, aluminum, fiberglass, and other commercially available materials. The material selection depends on factors such as: i) desired surface finish; ii) reusability; iii), construction ease. Wood products are the most widely used. Stay in place formwork is the one of the valuable CC technology contributions. Stay in place formwork is the one of the valuable CC technology contributions.

CC created concrete forms on the fly. These forms become an integral part of the concrete wall and are not removed after the concrete has cured. In CC formwork construction, a dense mortar is directly extruded by the CC machine to build the wall form; there is no need for separate formwork materials. This innovative approach can save labor and material costs and is easily integrated with current automation and robotics technologies.

For CC formwork, there are several desired mechanical properties such as bending and tensile strength, modulus of elasticity, and compressive strength. Compressive strength is usually the primary consideration in form design since the form itself is subjected to heavy vertical loads. Compressive strength of the CC form depends on cement mixture ratio, sand, and water. Through several trial and error experiments, a mortar mixture with characteristics suitable for CC construction was found. The mixture has the following composition:

- Plastic cement – 11.48 lb
- Sand – 13.18 lb
- Bentonite (a clay based plasticizer) – 0.8 lb
- Water – 7.03 lb

To verify its early compressive strength, three cylindrical test specimens, 2×4 inch (5×10 cm) were made and cured for 7 days in room temperature. Tests were conducted at USC's Civil Engineering structural testing laboratory and the results are shown in Table 9-1.

Table 9-1. Results of testing compressive strength

Cylinders	Compressive Strength (psi)
Specimen #1	2,786
Specimen #2	2,830
Specimen #3	2,606
Mean	2,741

These results indicate that the compressive strength of the three test specimens were both consistent and adequate and the mixture was suitable for use as a permanent structural component of the finished wall. For other applications, a chemical admixture might be an alternate means to obtaining the desired engineering properties.

9.1.3.7 CC nozzle geometry

One of the unique capabilities of the CC process is its superior surface forming capability. This is achieved using simple planar trowels which constrain the extruded flow in the vertical and horizontal directions. The friction force between the inner surface of the planar trowel and the outer surface of the extruded flow determines the final product surface finish. Too much friction drags the extruded material causing discontinuous flow, resulting in voids. The friction force is not always constant and can increase during a period of high extrusion pressure and excessive water seepage. Figure 9-21 shows how excessive friction force can cause the problems previously described, resulting in a defective surface finish.

A good surface finish however is not simply a matter of reducing the flow friction force. Delivering the right amount of material with consistency and precision is critical. It is difficult to control and deliver a highly viscous compressible paste through the CC nozzle assembly. A specialized CC nozzle assembly was designed. Figure 9-22 shows a schematic of the new CC nozzle assembly including the function of its components.

Rapid Prototyping: Theory and Practice 243

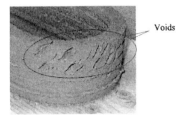

Figure 9-21. Excessive friction force causes some voids on extruded flow

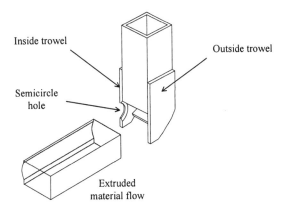

(a)

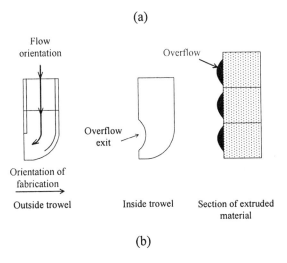

(b)

Figure 9-22. Specialized CC nozzle assemblies: (a) isometric view of the CC nozzle; (b) extrusion flow orientation and overflow control

The friction force decreases as the contact area of the outside trowel decreases. This reorients the material flow horizontally; inducing overflows which delivers excess material which creates the good surface finish of wall form. This was done without a flow feedback control device. Figure 9-23 shows the actual fabrication of a concrete wall form with the new CC extrusion nozzle. As the extrusion nozzle moves horizontally (X-axis), a layer is created with a smooth outer surface and a relatively rough inner surface. A smooth inner surface finish is unnecessary since concrete fills the internal volume. A rough inner surface may actually be desirable since it could help achieve a stronger bond with the freshly poured concrete.

(a) (b)

Figure 9-23. Extrusion flow control by new CC trowels: (a) outside trowel controls smooth surface finish; (b) overflow controls by inside trowel

9.1.3.8 Fabrication of vertical concrete formwork

The mortar mixture was prepared using power driven mixing paddles and loaded into the material carrying tank. The initial extrusion flow during start up is discarded until the flow is stabilized and the system starts its fabrication. Once an entire batch of the mortar mixture inside the material carrying tank is consumed, the CC system pauses until another batch is loaded. Extrusion then continues to complete the remaining concrete form. A batch of mortar is consumed in approximately 10 minutes and yields a concrete form approximately 2.5 inch (64mm) high.
Custom made form ties are manually inserted between layers while the CC machine is actively fabricating the concrete wall form. A tie is inserted at every 12 inches horizontally and 5 inches vertically. To complete a prototype section of full-scale concrete wall section nine batches were prepared. The final concrete form exceeded a height of 3 feet (60cm) and is shown in Figure 9-24.

(a) Concrete wall form made by the CC machine

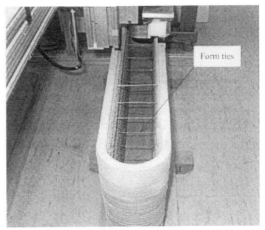

(b) Form ties

Figure 9-24. A concrete wall form fabricated by the CC machine: (a) desired span and height of wall form; (b) form ties are inserted according to beam theory analysis

9.1.3.9 Placing fresh concrete

Commercial concrete construction processing consists of mixing, transporting, and placing of the concrete. Truck-mixing is used in most construction fields. Concrete is transported with belt conveyers, buckets, and chutes depending on the construction field site and structure design. After the concrete is placed, it is compacted within the forms to remove lumps or voids. Hand tools or mechanical vibrators are used to guarantee a uniform dense structure.

In the CC process, placing concrete requires different procedures. A batch of concrete is poured to a height of 5 inch (13cm) and a second batch is poured on top of the first batch after one hour. A one hour delay batch is sufficient to control the lateral pressure of the concrete by allowing it to partially cure and harden. Even using one hour delay, it is possible to erect 10 foot high concrete wall in a day. The time delay needs to be adjusted depending on the concrete hardening rate. Accelerators might be a good solution when higher concrete placing rates are needed.

It is not necessary to wait until the concrete hardens completely since even partially cured concrete reduces the lateral pressure on the form. Figure 9-25 illustrates the CC concrete pouring process.

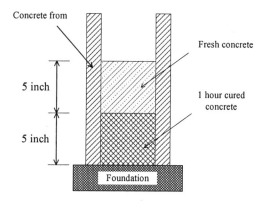

Figure 9-25. CC concrete placing procedures

The bottom cross-hatched section represents concrete which has been cured for one hour. The cured concrete produces minimal lateral pressure on the form but the exact value was difficult to quantify in our experiments. Exact data was not available in the available literature; therefore, a simple test was devised to validate the strength capability of the form. Figure 9-26 shows the test bed.

A batch of concrete was poured to a height of 5 inch (13 cm). The second batch was poured on top of the first batch after one hour without problems. This experiment was conducted without using any form ties. Structural concrete obtained from the actual construction field was used to fill the internal volume

Figure 9-26. Placing concrete in layer by layer without using form ties

9.1.3.10 Results

As described in subsection 9.1.3.9, concrete was manually poured into the extruded form in 13 cm incremental depths (one hour intervals) to a final height of 60 cm (2ft). Figure 9-27 shows the finished wall. The compressive strength of this wall will vary depending on the type of concrete chosen. Concrete pouring in this demonstration, however, has been independent of the extrusion forming process. With more experimentation, the filling process can be synchronized with the extrusion process. The coupling of these two processes will depend on many factors including extrusion rate, pour rate, curing time and strength requirements. In the next generation CC system, the mechanical assembly for continuous concrete pouring will be integrated into the CC extrusion nozzle assembly.

Figure 9-27. A concrete wall made by CC machine

9.1.3.11 Extraterrestrial construction

The ability to construct supportless structures is an ideal feature for building structures using in-situ materials. Hence we plan to explore the applicability of the CC technology for building habitats on the Moon and Mars. In the recent years there has been growing interest in the idea of using these planets as platforms for solar power generation, science, industrialization, exploration of our Solar System and beyond, and for human colonization. In particular, the moon has been suggested as the ideal location for solar power generation (and subsequent microwave transmission to earth via satellite relay stations).

Once solar power is available, it should be possible to adapt the current Contour Crafting technology to the lunar and other environments to use this power and in-situ resources to build various forms of infrastructures such as roads and buildings. The lunar regolith, for example, may be used as the construction material. Other researchers have shown that lunar regolith can be sintered using microwave to produce construction materials such as bricks. A Contour Crafting system may be developed to use microwave power to turn the lunar regolith into lava paste and extrude it through its nozzle to create various structures. Alternatively, lunar regolith may be premixed with a small amount of polymer powder and moderately heated to melt the polymer and then the mix be extruded by the CC nozzle to build green state (uncured) depositions in the desired forms. Post sintering of the deposition may then be done using microwave power. Understanding of the following is crucial for successful planetary construction using Contour Crafting: (a) the fluid dynamics and heat transfer characteristics of the extrudate under partial-gravity levels, (b) processes such as curing of the material under lunar or Martian environmental conditions, (c) structural properties of the end product as a function of gravity level, and (d) effects of extrudate material composition on the mechanical properties of the constructed structure.

One of the ultimate goals of the Human Exploration and Development of Space (HEDS) program of NASA is colonization, i.e., building habitats for long term occupancy by humans. The proposed approach has direct application to NASA's mission of exploration, with the ultimate goal of in-situ resource utilization for automated construction of habitats in non-terrestrial environments. The Contour Crafting technology is a very promising method for such construction[17].

9.1.4 Conclusion

Contour Crafting is the only layered fabrication technology which is suitable for large scale fabrication. CC is also capable of using a variety of materials with large aggregates and additives such as reinforcement fiber. The technology may be used for fabrication of large industrial objects such as molds and tooling for air frame industry. It can also be used for fabrication of boats and furniture frames.

Due to its speed and its ability to use in-situ materials, Contour Crafting has the potential for immediate application in low income housing and emergency shelter construction. Construction of luxury structures with exotic architectural designs involving complex curves and other geometries, which are expensive to build using manual approach, is another candidate application domain for CC. The environmental impact of CC is also noteworthy. According to various established statistics the construction industry accounts for a significant amount of various harmful emissions and construction activities generate an exorbitant amount of solid waste. Construction of a typical single-family home generates a waste stream of about 3 to 7 tons. In terms of resource consumption, more than 40% of all raw materials used globally are consumed in the construction industry Construction machines built for Contour Crafting may be fully electric and hence emission free. Because of its accurate additive fabrication approach Contour Crafting could result in little or no material waste. The CC method will be capable of completing the construction of an entire house in a matter of few hours (e.g., less than two days for a 200 m^2 two story building) instead of several months as commonly practiced. This speed of operation results in efficiency of construction logistics and management and hence favorably impacts the transportation system and environment.

There are numerous research tasks that need to be undertaken to bring the CC construction technology to commercial use. The activities reported in this article are the first few steps toward realization of actual full scale construction by Contour Crafting. Readers may obtain updated information on research progress and view video clips and animations of construction by CC at www.ContourCrafting.org

References

1. K. Behrokh, Innovative Rapid Prototyping Process Makes Large Sized, Smooth Surface Complex Shapes in a Wide Variety of Materials, Materials Technology, 13(1), 52-63 (1998).
2. L. Feng, S. Wei, and Y. Yongnian, Optimization with minimum process error for layered manufacturing fabrication, Rapid Prototyping Journal, 7(2), 73-81 (2001).
3. L. Feng, S. Wei, and Y. Yongnian, A mathematical description of layered manufacturing fabrication, Proceedings of Solid Freeform Fabrication Symposium, The University of Texas at Austin, Austin, TX, 139-46 (1999).
4. A. Bandyopadhyay, R.K. Panda, V.F. Janas, M.K. Agarwals, R. van Weeren, S.C. Danforth and A. Safari, Processing of Piezocomposites by Fused Deposition Technique, Proceedings of the 1996 10th IEEE International Symposium on Applications of Ferroelectrics, ISAF'96
5. N. Venkataraman, S. Rangarajan, M.J. Matthewson, B. Harper, A Safari, S.C. Danforth, G. Wu, N. Langrana, S. Guceri and A. Yardimci, Feedstock material property-process relationships in fused deposition of ceramics (FDC), Rapid Prototyping Journal, 6(4), 244-252 (2000).
6. K. Behrokh, R. Russel, H. Kwon, S. Bukkapatnam, Contour crafting-a layered fabrication technique, Special issue of IEEE robotics and Automation Magazine 8(3), 2001.
7. K. Behrokh, S. Bukkapatnam, K. Kwon, J. Saito, Experimental investigation of Contour Crafting using ceramic materials, Rapid Prototyping Journal, 7(1), 32-41 (2001).
8. K. Hongkyu, S. Bukkapatnam, B. Khoshnevis, J. Saito, Effect of orifice geometry on surface quality in Contour Crafting, Rapid Prototyping Journal, 8(3), 147-160 (2002).
9. K. Hongkyu, Experimental and Analysis of Contour Crafting (CC) Process using Uncured Ceramic Materials, PhD dissertation, Industrial and Systems engineering, University of Southern California (August 2002)

10 G. E. John., Member, ASCE, Hiroshi S., Construction Automation: Demands and Satisfiers in the United States and Japan, Journal of construction Engineering and Management, **122**(1), 147-151 (1996).
11 H. Dooil, Experimental Study of Full-Scale Wall Construction using Contour Crafting, Ph.D. dissertation, Industrial and Systems Engineering, University of Southern California (May, 2005).
12 C. Leslie and M. Nobuyasu, Construction Robots: The Search for New Building Technology in Japan, American Society of Civil Engineers (ASCE Press) (1998).
13 K. Behrokh, Automated Construction by Contour Crafting-Related Robotics and Information Technologies, Automation in Construction, **13**(1), 5-19 (2004).
14 A. Steven, Rapid Prototyping is Coming of Age, Mechanical Engineering, **117**(7), 62-68 (1995).
15 Sorptive Mineral Institute, Minerals and Ore, www.sorptive.org/minerals.html
16 Laguna Clay Company, Laguna Clay Catalog-Volume V, www.lagunaclay.com/volumev/product.htm
17 K. Behrokh, M. P. Bodiford, K. H. Burks, E. Ethridge, D. Tucker, W. Kim, H. Toutanji, M.R. Fiske, Lunar Contour Crafting – A novel technique for ISRU based habitat development, American Institute of Aeronautics and Astronautics Conference, Reno (January 2005).

Part III. Economic Analysis

Chapter 10

Strategic Justification of Rapid Prototyping Systems

Rakesh Narain[1] and Joseph Sarkis[2]
[1]*Department of Mechanical Engineering, Motilal Nehru National Institute of Technology, Allahabad-211004, India, narainr@rediffmail.com.* [2] *Clark University, Graduate School of Management, 950 Main Street, Worcester, Massachusetts, 01610, jsarkis@clarku.edu*

Abstract:

The consideration of whether to adopt rapid prototyping (RP) technology in general or a specific system in particular is not a trivial issue for most organizations. This technology has influences and has implications for a variety of intra-organizational functions and inter-organizational boundaries. The decision issues faced by these organizations include the balancing of needs across the organization and its partners, consideration of tangible and intangible factors, and the consideration of strategic and operational dimensions. In this chapter we introduce some of the categories of attributes and factors that organizations need to consider. The various factors are then evaluated using a multiattribute utility model called the analytical network process. An illustrative example provides insights into the execution of the technique. The technique is useful due to its capability to consider the many relationships and influences among the factors. It is also flexible enough to consider perceptual as well as more objective data and information when analyzing the problem situation.

Key words:

Strategic Technology Justification, Multiattribute Utility Theory, Analytical Network Process, Analytical Hierarchy Process, Decision Analysis, Technology Management, Investment Analysis, Multiple Criteria Decision Making.

10.1. Introduction

Time to market is a crucial success factor for most organizations. The winners in today's competitive marketplace are those companies that can bring innovative high value products & services to the customer ahead of their competitors. Rapid Prototyping (RP) technology enables an organization to acquire such a capability[12]

RP is a term which includes a range of new technologies for producing accurate parts directly from 3D-CAD models, without the use of tooling, in a few hours with minimal human intervention. RP uses state-of-the-art laser technology, positioning systems, material and process technologies in various processes.

Ever since the birth of RP technology, around mid 1980's there has been tremendous technical advancement and global proliferation of RP technology. Starting with the single commercially available RP system known as Stereolithography manufactured by 3D Systems, today we have a score of RP systems available. These include such technologies as: Solid Ground Curing (SGC), Solid Object Ultraviolet Laser Plotter (SOUP), Laminated Object Manufacturing (LOM), Fused Deposition Modeling (FDM), Selective Laser Sintering (SLS), 3D Printing etc. A detailed classification of RP has been given by Kruth[11] and Pham and Gault[14].

As per an estimate, by the year 2002, some 10,000 RP systems were to be installed through out the world generating global revenue of RP of over $700 million. According to the Wohlers report[19], the RP industry, which includes product sales and services worldwide, grew 9.2% to $ 528.9 million in 2003, up from $ 484.5 million in 2002. The industry is expected to grow to $ 586 million in 2004 and $ 655 million in 2005. A difficulty and variation in the estimates is due to definitions of RP technology, but clearly industry is spending hundreds of millions of dollars on this technology with a continuous increase as product markets continue to customize.

Initially the technology was used as a product development tool, mainly to evaluate the ease of assembly and manufacture of designed products. But over the years the area of application has widened into other fields including rapid tooling, medical modeling, automotives, aerospace parts, building construction, jewelry, home appliances, to name a few.[8]

RP Technologies are generally known as an additive fabrication process as they build parts through layer-by-layer addition of material, in contrast to the subtractive fabrication process (such as lathe, milling, drilling, computer numerical wire cuts, and electronic data management) where the part is built by removal of material from a bigger stock. The subtractive fabrication processes offer several advantages compared to the additive fabrication i.e. R.P. process, e.g. Better accuracy and surface finish, efficient bulk fabrication, material versatility, structural integrity and large build envelops[4]. Still the R.P. processes, compared to the subtractive processes, offer a whole new set of advantages. These include, offering unlimited geometrical complexity, engineered milling structure, unattended operations and waste less fabrication. The raw materials used in RP come in different forms such as liquid resins, powder, plastic filaments and paper. Most of the RP processes are classified as layer additive processes and layer subtractive process. In layer additive process thin layers of material are gradually added to the concept model or prototype unit it is completed. The layer-subtractive process differs from layer additive processes in that it builds a concept model or prototype and removes the excess material at the same time. This method of building is generally considered a faster RP method.

RP is considered as a time compression technique which offers numerous benefits. Several case studies are reported in the literature showing dramatic reduction in product development time and cost, and also an increase in revenue due to early, introduction of products to market. For example, a cylinder head flow box that normally took 320 hours to fabricate at a cost of $10,000 was produced by RP in 80 hours. It was estimated that one automotive manufacturer saved $10 million in a two-year period using various forms of RP[10]. Despite all these potential benefits, the acceptance of RP technology by industry in product development cycle has been frustratingly slow. Though there are many reasons for it[20] one of the most important one is the high initial capital investment required for the purchase of RP systems, and the inability to justify such investments using historical cost accounting justification methods. As a result of this difficulty in investment most small and medium sized companies cannot afford its purchase, or shy away from high prices charged by engineering service bureaus for rapid prototyping[22]. The uncertainty and high risk on the rate of investment and capital return have further built barriers between the technology providers and technology users.

To promote this emerging technology in industrial applications and for its wider adoption, there is a compelling need to develop procedures to justify the purchase of such expensive RP systems within a broader business case perspective. To that extent we provide a model to be able to complete this analysis. The model is an extension of the analytical hierarchy process (AHP) called the analytical network process (ANP). We shall provide an illustrative example of how this technique may be applied using various generic factors for evaluation.

We begin by initially considering various metrics and factors to consider in the evaluation of these systems.

10.2. Investment Justification Factors

There are numerous factors that should be considered when dealing with technology that can significantly impact the strategic and operational activities and functions of any organization. We will categorize them into four primary categories for our decision analysis. These are strategic and operational organizational factors, system cost issues, and other system characteristics.

10.2.1 Strategic and Operational Benefits

While cost elements can be easily identified, and are identified for the reader below, difficulty is experienced in quantifying the numerous non-cost oriented benefits of RP. For example, companies are attaching greater importance to the need to differentiate their product and service offerings in order to remain competitive, RP technology aids in this process, yet its explicit inclusion in the justification process does not typically occur. As we know, from a strategic perspective firms are:
- adding new technologies to their products to differentiate;
- forming alliances with their customers;
- adding service features to their manufactured product offerings;
- reducing time to market for new products;
- reducing the number of suppliers and forming longer-term relationships and alliances with those that remain;
- expanding their product range; and
- reducing their cost base to become the lowest cost producer.

Given these characteristics, we can define two general categorizations of RP benefits:

    A. The Direct (or operationally oriented) Benefits e.g. are: It
        (i)   Reduces design phase cycle time and costs
        (ii)   Reduces the potential for expensive design errors
        (iii)   Reduces tooling costs on short-run parts
        (iv)   Reduces time to production & market.

    B. The Indirect (strategically oriented) benefits examples include:
      (i) Help enter new markets
      (ii)Increases market share by introducing more innovative products that customers will value

(iii) Offering more customized products with no time or cost penalty

10.2.2 Cost

Numerous cost elements exist in this system[3,21], as is true in most strategic systems an organization evaluates. Some of the costs are easily determined while others may require significantly more effort. A numerical listing is provided here since most are self explanatory:

1. Cost of RP System
2. Cost of Computers and Softwares (For solid modeling work)
3. Cost of special and proprietary Softwares (For creating of support structures & generation of slicing files)
4. Cost of Software upgrades
5. Material Cost
6. Labor Cost (Cost of making solid models, setting up cost of machine, cleaning the supports, the post process curing, polishing, smoothing etc)
7. Annual maintenance Cost (Approximately 10-15% of the purchase price of equipment)
8. Cost of Disposal of hazardous material
9. Cost of scrap (Due to bad builds, or damaged parts)
10. Cost of waste material (Due to partial curing, or incomplete nesting)
11. Facility Rehabilitation cost (to house RP machine)
12. Training Cost
13. Delivery Cost
14. Cost of uninterruptible power supplies
15. Cost of obsolescence
16. Other Costs (such as laser replacement cost etc.)

There are a variety of methods to categorize these costs. For the sake of keeping the amount of analysis and evaluation to a minimum in our illustrative example, we can categorize the costs into five major areas and their benefits/costs: Initial Systems Costs, Labor Costs, Material Costs, Operational Costs, and Maintenance Costs (we could have also categorized them into fixed and variable costs). These costs can be traced to both the system itself and the costs associated with the function and activities the system supports (e.g. scrap costs and changes).

10.2.3 Systems Characteristics and Factors

In making an evaluation or justification, the actual characteristics of the system, its hardware, software, etc., need to be evaluated. Again, many of these characteristics are not easy to quantify and have numerous intangible dimensions. We shall now provide a quick overview of some of the major characteristic elements and factors.

10.2.3.1 Inter- and Intra-Firm Adaptability

RP packages require sophisticated hardware and software systems that are often costly, notwithstanding the drop in computer prices. Not only should these new systems match the current hardware and software (mainframes, legacy systems, etc.), but they should also, more importantly, match the sophistication levels of would-be users. Finally, since no packaged software can meet all the requirements of an enterprise, it may often be necessary to buy third-party tools.

Since the supply chain extends beyond the boundaries of an organization, an RP system must be able to communicate with its counterparts at the supplier and customer sites. Such a communication may be warranted in the context of the growing trend towards e-commerce.

10.2.3.2 Platform Neutrality and Interoperability

The architecture of RP systems must be platform flexible. Related to some of the adaptability characteristics, these factors are usually at the initial stages and may require special initial importance.

10.2.3.3 Scalability

The size of a business is hardly static and fluctuates with time. The performance of the RP system must scale well with business size; an absence of this can contribute to a failure.

10.2.3.4 Reliability

We can measure the reliability of information systems' components in a number of ways. In the reliability literature, especially software reliability, the commonly used reliability measures are: number of faults in a fixed time interval; and time between two faults. Faults in the system can occur due to disconnected lines, system crashes, or an inappropriate, unreliable or wrong response. Such faults not only decrease productivity, but also diminish the confidence in the system.

10.2.3.5 Ease of Use

Employees cannot afford to spend a lot of time to learn new software. The software has to be intuitive (e.g., GUI) and reflect the mental picture of the business objects, associations, and business activities with which users are familiar. Ease of use can affect the success and cost (e.g. training) of the RP system.

10.2.3.6 Customer Support

Companies are beginning to realize the importance of customer support in the marketing of their software products. Training, education, and reference materials must be available on demand, and responses to customer's questions must be done in a timely manner. This is more the case with RP systems whose complexities necessitate excellent customer (vendor supplied) support to ensure that employees use the system and also perceive it to be of value.

10.3. Issues and Models for Evaluation and Justification of RP

All the benefits of RP are difficult to accurately forecast. This makes the calculation of traditional approaches such as return on investment (ROI) difficult. This has led many to justify investment in RP on subjective factors such as emotions, personalities and personnel challenges than on objective data[7]. This may lead to erroneous judgment.

RP has the potential to be recognized as a tool of tomorrow which must be learned today, to assure organizations of a share in tomorrow's prosperity (i.e. remain a sustainable enterprise). The importance justification for RP is obvious. Strategically we organizations also need to ask, 'how effectively can we compete with our competitors a few years hence, if we do not use RP today?'

Therefore careful analysis of each element of cost and benefits is required so that a model can be developed after including each of these costs/benefits in the investment appraisal and suitably accounting for the elements of risk also. We seek to now introduce a model that incorporates a number of these concerns and presents a way to help objectify some of the subjectivity of the decision analysis.

10.3.1 The Analytical Network Process (ANP) Methodology

As we have stated strategic decisions and evaluation should seek to incorporate tangible, intangible, strategic and operational factors into any analysis. ANP a generalized form of the more popular multiattribute decision making tool the Analytical Hierarchy Process (AHP)[15], can help to perform this task.

Firstly, AHP has been a popular approach for multiattribute evaluation of technologies and other strategic decisions.[1,13] It is a robust technique that allows managers to determine preferences of criteria for selection purposes, quantifying those preferences then aggregating them across diverse criteria. It is a relatively easy approach to understand and apply. AHP has been applied for the selection of Rapid Prototyping technologies[2]. However, it has certain difficulties that arise when developing an appropriate framework that is acceptable and useful for management. First, AHP only considers one-way hierarchical relationships among factors. This simplistic assumption does not consider the many possible relationships among the groups of factors or those within them. For example, for selection of a project, a decision-maker may categorize factors into cost, quality, and flexibility. A project may be rated on each of these factors separately and aggregated to arrive at an overall score, which is essentially what AHP does. Yet, AHP does not explicitly consider the interactions among the various factors (e.g. cost and quality may impact flexibility). ANP is capable of incorporating this and many other inter-relationships of factors into the decision model. AHP is also guilty of the possibility of 'rank reversal'[5]. This problem is mitigated with the ANP method[17], making it a potentially more accurate and useful decision support tool for strategic evaluation, justification and selection. Still, the ANP approach may be disadvantageous due to its increasing complexity as the number of factors and relationships increase.

10.3.1.1 Step 1: Setting up the Network Decision Hierarchy

There are five major steps in applying the ANP technique. The first step is to develop a network of factors influencing the final decision- this is known as the ANP network decision hierarchy. In this network decision hierarchy there is no clearly discernable hierarchy, but doe typically have a general objective itself with various factors, dimensions and alternatives that need to be evaluated. Instead of hierarchical levels, the factors are grouped into clusters that may have numerous controlling relationships. The ANP network decision model is usually finalized after a considerable amount of discussion at numerous levels within and between organizations. This discussion is quite valuable because structuring the decision problem helps managers and the decision maker see a logical "big picture" and the issues involved in a complex decision.

Within ANP, AHP's standard hierarchical decision approach is altered. This alteration includes a number of interdependent relationships among the various factors and evaluation criteria that did not exist in AHP. We show these various relationships and overall view for and RP technology network evaluation model in Figure 10-1. In this figure is the objective of strategically evaluating the RP technology called the decision (DEC). In this network decision hierarchy we include a variety of clusters, a strategic cluster (from our definitions of factors earlier) called the Strategic Benefits (SB) cluster and the planning horizon cluster (PlH). Instead of hierarchical levels, as in AHP, it is recommended that the use of the term clusters should be used to signify a grouping of factors[16]. The term 'level' would imply a hierarchical top-down control relationship among factors, whereas clusters may have many directional relationships. Each cluster is composed of a number of factors.

Notice the interrelationship, interdependencies of these two clusters. The SB clusters influences and is influenced by the planning horizon factors cluster. The two-way arrows among the levels identify these interdependent relationships. For example, one direction would determine which of the strategic benefits is most important over a 'short-term' planning horizon, which factor over the 'long-term'. The other directional relationship would question a specific strategic benefit's importance in the short-term and the long-term.

The RP alternatives (ALT) are then each evaluated on various tactical and operational factors. These tactical and operation factors are detailed include the direct benefits (operational metrics) defined earlier such as design time, design errors and time to production. System based metrics were also defined and are included in another cluster. The final set of metrics is cost based metrics and interrelates to system based metrics.

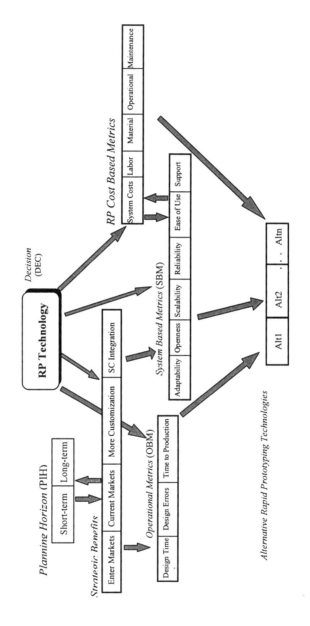

Figure 10.1. Network Decision Hierarchy for ANP evaluation of RP Technology Justification Decision.

10.3.1.1.1 The Planning Horizon

We have described most of the clusters included in our evaluation in a previous section on discussion of factors for evaluation. We did not discuss the planning horizon and will now provide an overview. This cluster incorporates the dynamic and competitive nature of any organization. Incorporating the planning horizon into a decision model will integrate evolving strategies and directions of the organization, which might be critical for strategic decisions. For example, managerial decision-makers want to make sure that long-term issues are covered from a strategic perspective, but short-term risks may cause a barrier such that the long-term horizon may never be reached. Thus, incorporating both short-term and long-term considerations is important.

In addition, there are a number of other temporally based factors that can be used within this cluster. Time period measures (years, quarters, etc.) or using the product life cycle stages, are alternative approaches for incorporating temporal or dynamic characteristics of strategic decision environments. Incorporating competitive environmental dynamism addresses how factors may have differing (evolving) importance over time. For the sake of maintaining some simplicity in the model, we apply the dichotomous categorization of short-term versus long-term planning horizons. These two factors are depicted in the detailed network decision hierarchy shown in Figure 10-1.

To elicit information from these relationships a decision-maker may be asked the following pairwise comparison question to help determine the relative importance of a strategic performance metric: "How much more important is cost when compared to flexibility over the long-term." The answer to questions such as these will help us complete the pairwise comparison matrix, which is further detailed below in the various elicitation and execution steps of ANP. Similar questions may be asked of all relationships.

10.3.1.2 Step 2: Pairwise Comparisons

The second step is to elicit pairwise comparisons (PWCs) between the factors, using inputs from users/managers. The ANP approach will then require analysts to systematically elicit inputs by asking users to evaluate the relative importance of one factor when compared to another factor--pairwise comparisons--with respect to a third controlling factor. The controlling factor may be at the same or different levels (clusters) in the ANP model. For each controlling factor, the corresponding set of pairwise comparisons can be represented in matrix form. Saaty[15] suggests that the values assigned to the comparisons of the factors be made in the range 1/9 to 9. A 9 indicates that one factor is *extremely more important* than the other, a 1/9 indicates that one factor is *extremely less important* than the other, and a 1 indicates equal importance.

Table 10-1 shows a pairwise comparison matrix for the strategic metrics factors based on the objective (the controlling factor). An example pairwise comparison question for this matrix would have been: "How much more important is SB1 (entering new markets, in this example) than SB2 (expanding current markets) in when evaluating a RP technology alternative?"

Table 10.1. Example Pairwise Comparison Matrix and resulting Relative Importance Weight for Strategic Metrics (Cluster with respect to Decision).

DEC	SB1	SB2	SB3	SB4	$w^{j,k_{DEC}}$
SB1	1	3	2	4	0.462*
SB2	0.333	1	0.5	2	0.156
SB3	0.5	2	1	4	0.294
SB4	0.25	0.5	0.25	1	0.088

* numbers in the shaded column are obtained from equation 1.

10.3.1.3 Step 3: Calculate Relative Importance Weights

From each pairwise comparison matrix obtained in Step 2, we calculate the relative ranking (importance weights) of factors with respect to the corresponding controlling factor. An eigenvalue problem needs to be solved to estimate these relative importance weights.

The eigenvalue solution provides us with the local priority vector w which is computed as the unique solution to:

$$Aw = \lambda_{max} w, \qquad (1)$$

where λ_{max} is the largest eigenvalue of pairwise comparison matrix A. The solution for the eigenvalues and relative importance weights (w) can be determined utilizing various software tools, one is the Web HIPRE3+ software located at www.hipre.hut.fi. For this example, from Table 10-1, w = {**0.462, 0.156, 0.294, 0.088**}. This essentially shows that SB1 is the most relatively important factor for our decision and for this organization. SB4 is the least important factor. These factors may be adjusted due to the interrelationships that exist and the next two steps will be used to integrate

Rapid Prototyping: Theory and Practice

these various relationships.

10.3.1.4   Step 4: Form a Supermatrix

In this step a two-dimensional matrix composed from the relative-importance-weight vectors found in step 3 is formed. This matrix is called a supermatrix. For example the relative importance weights from our illustrative Step 3 are shown as bold numbers in the supermatrix in Table 10-2. The row elements represent the factors relating to the control factor in that column. For example, the bold numbers represent the relative importance of SB1-SB4 on the DEC objective. After this initial formation the supermatrix needs to be made 'column stochastic' by making sure the summation of all the elements in a column sum to 1. There are two ways to complete this substep. First is to assign relative importances to various clusters of elements on a given control factor (column). For our example, we will do this for the first column by multiplying .4 to the SB factors, .2 to the OBM (operational metrics) factors, .2 to the SBM (system based metrics) and .2 to the RBM (RP cost based metrics) values. The other approach which we use for each of the other columns to normalize them to one is to divide each element in that column by the summation of that column.

10.3.1.5   Step 5: Arrive at a Converged set of Weights

The interdependencies of the weights can now be determined through a Markovian based process that arrives at a final set of converged weights. This can be completed by raising the normalized supermatrix from Step 4 by a large power to calculate the converged ("stable") weights for the alternatives. We set the goal of converging the weights to at least the 10^{-4} level. Thus, we essentially apply expression (2) to this step.

$$\mathbf{W}_F = \lim_{p \to \infty} (\mathbf{W}_I)^p \qquad (2)$$

Were $\mathbf{W}_F$ and $\mathbf{W}_I$ represent the final and initial supermatrices, respectively. In our example we consider three alternatives (ALT1 to ALT3).

Table 10-2. Initial Supermatrix.

	DEC	PH1	PH2	SB1	SB2	SB3	SB4	OBM1	OBM2	OBM3	SBM1	SBM2	SBM3	SBM4	SBM5	SBM6	RBM1	RBM2	RBM3	RBM4	RBM5	ALT1	ALT2	ALT3
DEC	0	0	0	0	0	0	0	0	0	0	0	0	0	0	0	0	0	0	0	0	0	0	0	0
PH1	0	0	0	0.400	0.350	0.700	0.600	0	0	0	0	0	0	0	0	0	0	0	0	0	0	0	0	0
PH2	0	0	0	0.600	0.650	0.300	0.400	0	0	0	0	0	0	0	0	0	0	0	0	0	0	0	0	0
SB1	**0.462**	0.109	0.371	0	0	0	0	0	0	0	0	0	0	0	0	0	0	0	0	0	0	0	0	0
SB2	**0.156**	0.109	0.41	0	0	0	0	0	0	0	0	0	0	0	0	0	0	0	0	0	0	0	0	0
SB3	**0.294**	0.433	0.126	0	0	0	0	0	0	0	0	0	0	0	0	0	0	0	0	0	0	0	0	0
SB4	**0.088**	0.349	0.093	0	0	0	0	0	0	0	0	0	0	0	0	0	0	0	0	0	0	0	0	0
OBM1	0.429	0	0	0.714	0.6	0.455	.2	0	0	0	0	0	0	0	0	0	0	0	0	0	0	0	0	0
OBM2	0.143	0	0	0.143	0.2	0.455	0.6	0	0	0	0	0	0	0	0	0	0	0	0	0	0	0	0	0
OBM3	0.429	0	0	0.143	0.2	0.091	0.2	0	0	0	0	0	0	0	0	0	0	0	0	0	0	0	0	0
SBM1	0.32	0	0	0.32	0.169	0.109	0.063	0	0	0	0	0	0	0	0	0	0	0	0	0	0	0	0	0
SBM2	0.149	0	0	0.149	0.098	0.218	0.105	0	0	0	0	0	0	0	0	0	0	0	0	0	0	0	0	0
SBM3	0.073	0	0	0.073	0.288	0.218	0.3	0	0	0	0	0	0	0	0	0	0	0	0	0	0	0	0	0
SBM4	0.131	0	0	0.131	0.288	0.252	0.174	0	0	0	0	0	0	0	0	0	0	0	0	0	0	0	0	0
SBM5	0.266	0	0	0.266	0.059	0.102	0.058	0	0	0	0	0	0	0	0	0	0	0	0	0	0	0	0	0
SBM6	0.061	0	0	0.061	0.098	0.101	0.3	0	0	0	0	0	0	0	0	0	0	0	0	0	0	0	0	0
RBM1	.089	0	0	0	0	0	0	0	0	0	0.320	0.244	0.273	0.243	0.273	0.186	0	0	0	0	0	0	0	0
RBM2	.294	0	0	0	0	0	0	0	0	0	0.058	0.186	0.093	0.183	0.093	0.209	0	0	0	0	0	0	0	0
RBM3	.162	0	0	0	0	0	0	0	0	0	0.144	0.198	0.281	0.192	0.280	0.244	0	0	0	0	0	0	0	0
RBM4	.162	0	0	0	0	0	0	0	0	0	0.246	0.163	0.145	0.158	0.145	0.163	0	0	0	0	0	0	0	0
RBM5	.294	0	0	0	0	0	0	0	0	0	0.231	0.209	0.209	0.224	0.208	0.198	0	0	0	0	0	0	0	0
ALT1	0	0	0	0	0	0	0	0.290	0.429	0.429	0.143	0.429	0.429	0.333	0.333	0.429	0.333	0.637	0.081	0.081	0.258	1	0	0
ALT2	0	0	0	0	0	0	0	0.655	0.143	0.429	0.143	0.429	0.143	0.333	0.333	0.429	0.333	0.258	0.188	0.188	0.105	0	1	0
ALT3	0	0	0	0	0	0	0	0.055	0.429	0.143	0.714	0.143	0.429	0.333	0.333	0.143	0.333	0.105	0.731	0.731	0.637	0	0	1

Since this is a generic example we do not specify the characteristics of these technologies. The resultant converged set of weights for the ALT factors under the DEC column, is W_F (ALT1, ALT2, ALT3t) = {.330, .342, .328}. Note that in this situation we would say that RP alternative number 2 is the highest ranking. Clearly an evaluation of the robustness of this solution can be completed by varying some weights and seeing how much the final solution changes in a 'what-if' analysis.

10.4. Summary and Discussion

RP technology seems to possess immense potentials for strategic and operational improvements within a wide variety of organizations and industries. To ensure a rapid rate of diffusion of this technology several steps are being taken namely:
1. Increasing the range of materials (and of course the size of the built model) that the RP machines can process.
2. Extending the range of applications of RP to more diverse areas such as from fabrication of miniature parts (some as tiny as a red blood cell), to modeling of historical monuments (to preserve its architectural details), to producing 3D models of large asteroids.
3. Bringing down the cost of machine by way of more development, and at the same time increasing the awareness among people about this technology.

Nevertheless, even with these advances, a fear still looms large in the minds of the users regarding the return they would be getting from the huge investment made in the purchase of RP technology, lest it should prove to be a white elephant. In the absence of any tried and tested method to justify their investment, organizations prefer to continue taking help of the service bureaus for their RP needs which at times may be more expensive than having in-house RP systems[18]. It is here that the understanding of the benefits (particularly the strategic benefits) coming out from RP technologies need to be fully understood. Then only all the benefits can be included in the analysis and a reasonable decision arrived at.

In this chapter we have presented a methodological approach to help evaluate the relatively strategic decision of justifying an RP technology. Within this process we have generated a number of factors and clusters for evaluation. Some of these factors are general business and operational issues that may be influenced by the type of RP technology selected. Other factor groupings are specific to characteristics of the RP technology itself. These clusters of factors included both tangible and intangible, as well as strategic, operational, technical, and business dimensions. To be able to effectively evaluate all these items we recommended the application of the ANP approach.

The ANP approach may be beneficial in this circumstance but decision analysts and managers need to be aware of the practical application of such a technique. First there are issues of how the data is gathered and aggregated. The acquisition of data could either be by an individual or group. Clearly, the perceptions of different individuals within an organization would vary depending on background and expertise. There are issues of power and political pressures that might also exist. In terms of getting responses from various individuals requires some aggregation and/or consensus building that must be accomplished. Secondly, the determination of the decision hierarchy in terms of the relationships and interrelationships that exist amongst the factors and clusters needs to be investigated. Additional and/or fewer relationships may exist and thus agreement among them should exist. The factors themselves and how many to include is something that might be corporation or industry specific depending on the strategies and operations of those specific organizations. Finally, the results may show a situation where the discrimination and rankings of the evaluated RP tools is very distinct. Calling for and doing a sensitivity analysis is something that is not as clearcut in ANP (as compared to AHP). The sensitivity of such a model and its robustness would need to be investigated.

Many of these issues can be studied and evaluated in a real world application of the model on this tool.

References

1. G. Brosoglu, and T. Tazgac, An application of the analytic hierarchy process to the supplier selection problem, Production and Inventory Management Journal, **38**(1) 14-21(1997).
2. M. Braglia and A. Petroni, A Management-Support Technique for the Selection of Rapid Prototyping Technologies, Journal of Industrial Technology, **15**(4), 2-6(1999).
3. R. Brown, and K. W. Stier, Selecting Rapid Prototyping Systems'. Industrial Technology, **18** (1), 2-8(2002).
4. M. Burns, Automated Fabrication: Improving Productivity in Manufacturing. PTR Prentice Hall, Englewood Cliffs, New Jersey (1993).
5. J.S. Dyer, Remarks on the Analytic Hierarchy Process, Management Science **36** (3), 249-268(1990).

6. K. Fritz, and F. Carsten, , 'Rapid Prototyping and Rapid Tooling'. Fraunhofer Institute Produktionstechnologie IPT, Steinbachstrabe (2002).
7. http://www.ipt.fraunhofer.de.
8. T. Grimm, Should you invest in rapid prototyping, Design Fax. June, [dfx/incl/head.asp] [dfx/incl/99dfx.htm] (2000).
9. C. C. Kai, and L. K Fai, Rapid Prototyping- Principles and Applications in Manufacturing. John Wiley and Sons (Asia) Pte Ltd. Singapore (1997).
10. P. R. Kleindorfer and F. Y Partovi, Integrating Manufacturing Strategy and Technology Choice, European Journal of Operational Research, **47**(1), 214-224(1990).
11. S. Krar and A. Gill, 'Justify your ROI, Advanced Manufacturing Magazine-Tomorrow's ideas at work today (2000) www.advancedmanufacturing.com/January 02/exploringamt.htm
12. J. P Kruth, Material incress manufacturing by rapid prototyping technologies, CIRP Annals, **40**(2), 603-614 (1991).
13. C. J. Murray, Design Cycles Shrink, In Design News, Available www.manufacturing.net/magazine/dn/archives/1998/dn0706.98/features3.html (1998).
14. R. Narasimhan, An analytical approach to supplier selection, Journal of Purchasing and Materials Management, **19**(1), 27-32 (1983).
15. D. T. Pham, and R. S Gault, A comparison of rapid prototyping technologies, International Journal of Machine Tools and Manufacture, **38**(1), 1257-1287 (1998).
16. T.L. Saaty, The Analytic Hierarchy Process, New York: McGraw Hill (1980).
17. T.L. Saaty, The Analytic Network Process, New York: McGraw Hill (1996).
18. S. Schenkerman, Avoiding Rank Reversal in AHP Decision-Support Models, European Journal of Operational Research, **74**(1) 407-419 (1994).
19. T. Wohlers, RP brings 3D printing to the Design office, Computer Aided Engineering, **15**(8), 5 (1996).
20. T. Wohlers, RP Industry Reverses Downward Trend, Web Log 43, 30 May, 2004. mhtml:file://G:\New Folder\wohlers Associates-Wohlers Talk.mht (2004).
21. T. Wohlers, and T. Grimm, Obstacles to RP's Growth, In Perspectives a monthly column in Time Compression Technologies. October issue (2001).
22. T. Wohlers, and T. Grimm, The Real Cost of RP, In Perspectives a monthly coloumn in Time Compression Technologies. Vol, March/April, file://A:\RPF\perspectives-mar 2002.htm(2002).

23. G. Zhang, M. Richardson, and R. Surana, Economic Evaluation of Rapid Prototyping in the development of New Products, technical Research Report T.R. 97-35, and Sponsored by the National Science Foundation Engineering research Centre Program, The University of Maryland, Harvard University and Industry (1997).

Chapter 11

SELECTION OF RAPID MANUFACTURING TECHNOLOGIES UNDER EPISTEMIC UNCERTAINTY

Jamal O. Wilson and David Rosen
Systems Realizations Laboratory, The G.W. Woodruff School of Mechanical Engineering, Georgia Institute of Technology, Atlanta, GA 30332-0405

Abstract:

Rapid Prototyping (RP) is the process of building three-dimensional objects, in layers, using additive manufacturing. Rapid Manufacturing (RM) is the use of RP technologies to manufacture end-use, or finished, products. At small lot sizes, such as with customized products, traditional manufacturing technologies become infeasible due to the high costs of tooling and setup. RM offers the opportunity to produce these customized products economically. Coupled with the customization opportunities afforded by RM is a certain degree of uncertainty. This uncertainty is mainly attributed to the lack of information known about what the customer's specific requirements and preferences are at the time of production. In this paper, we present an overall method for selection of a RM technology under the geometric uncertainty inherent to mass customization. Specifically, we define the types of uncertainty inherent to RM (epistemic), propose a method to account for this uncertainty in a selection process (interval analysis), and propose a method to select a technology under uncertainty (Hurwicz selection criterion). We illustrate our method with examples on the selection of an RM technology to produce custom caster wheels and custom hearing aid shells.

Key words:

Rapid Manufacturing; Selection; Epistemic Uncertainty; Decision Support Problem Technique.

11.1 Introduction

Rapid Prototyping (RP) is the process of building three-dimensional objects, in layers, using additive manufacturing. This process is done quickly, relative to other "one-off" manufacturing techniques. Companies of all sizes rely on RP in an effort to reduce time to market, improve quality, and reduce costs[1].
Rapid Manufacturing (RM) is the use of the RP technologies to manufacture end-use products, or finished parts. Recent studies have shown that companies have a strong interest in using RP to produce customized products. Some examples include Siemens and Phonak (hearing aid shells), Boeing's Rocketdyne (manufacturing hundreds of parts for International Space Station and the space shuttle fleet, F-18 fighter jets, etc.), etc[1]. There is also strong interest by the biomedical field in these types of technologies.
Given RM's relatively recent introduction, there is still a lot of skepticism surrounding these technologies. Some particular areas of concern are the part cost, build time, and production quality of the parts produced using RM, compared to that of traditional manufacturing technologies. Equally important, how does one select one of these technologies out of over 34 worldwide manufacturers of these machines? We believe that a systematic method for selection of the RM technologies will be critical to its emergence in the manufacturing enterprises of tomorrow.
One of RM's main advantages is its ability to produce customized parts. At large lot sizes, conventional manufacturing technologies have proven to be the most economical. At small lot sizes (such as the case for customized parts), because of the high cost of tooling and setup, conventional manufacturing technologies become infeasible. This is where RM is a key. RM offers the ability to produce large amounts of highly customized parts at a relatively fast pace. This customization ability introduces considerable

amount of uncertainty about what the customer wants and will choose. In this case, uncertainty lies in the geometric shape of the customized parts. In this chapter, we propose a method to aid the designer in accounting for this type of uncertainty (geometric uncertainty due to customization) in the selection process.

There have been several methods previously developed to account for uncertainty in the selection process. These methods can be grouped by the way uncertainty is represented: stochastic methods[2-4], fuzzy methods[5,6], and interval methods[7]. Reddy et al.[7] have applied exact interval arithmetic to represent uncertainty in the context of material selection for aircraft tires. In our chapter, we focus on the types of uncertainty involved in the selection of a RM technology. Although we represent uncertainty using intervals, we acknowledge that not all types of uncertainty in the selection process should be modeled using interval analysis. We also present a systematic method of selection of technologies with overlapping performance intervals.

11.2 UNCERTAINTY AND ITS REPRESENTATIONS

Before being able to account for uncertainty, we must first understand and classify the type of uncertainty that we are accounting for. Uncertainty can be divided into 2 distinct types: aleatory and epistemic. Aleatory uncertainty can be considered as irreducible or inherent uncertainty[8]. Epistemic uncertainty can be considered subjective or reducible uncertainty that stems from lack of knowledge or data[8]. Aleatory uncertainty can be easily quantified and represented by a probability density function (PDF), while epistemic uncertainty lacks the information needed for representation with a complete PDF. This is mainly due to the fact that with a PDF, you are predicting the likelihood of an event to occur. With epistemic uncertainty, this likelihood cannot be quantified due to lack of information or data.

Specifically, for RM, the geometric uncertainty (due to customization) is the main source of uncertainty. Because the designer lacks the information about what the customer will actually choose at the time of production, there is geometric uncertainty about how the part will be shaped. In this case, geometric uncertainty due to customized products is considered natural variation. RM allows the customer to choose from a finite/infinite range of customization. For simplicity, geometric uncertainty is the only source of uncertainty that will be considered in this chapter. Although the other uncertainties are important, they can be considered marginal when compared to the geometric uncertainty associated with customization. Thus, only epistemic uncertainty associated with the natural variation of customization is considered in this chapter. Natural variation is uncertainty that occurs in systems that change in ways that are difficult to predict.

The two most prominent ways of representing natural variation are probability theory and interval analysis. In the probabilistic approach of representing epistemic uncertainty, under Laplace's Principle of Insufficient Reason, uncertainty is modeled with a uniform distribution across the range[9,10]. When using the interval representation of epistemic uncertainty, uncertainty is modeled with a closed interval bounded by z_{min} and z_{max} (i.e., $Z \in [z_{min}, z_{max}]$). When epistemic uncertainty is modeled using interval numbers, the design equations are converted to intervals[11]. These intervals are then propagated through using interval arithmetic[12]. This process results in a bounded interval that represents the uncertainty in the results. It should be noted that interval operations must be carried through all computations to ensure the results accurately reflect the uncertainty in the results. There have been many arguments in favor of and against the use of interval analysis[13] and probability theory[14] when epistemic uncertainty is present. In our case, where there is limited information known about the geometric uncertainty involved, we believe the use of the interval representation of epistemic uncertainty is justified.

Now that the interval representation of epistemic uncertainty has been selected, the question now becomes how does one select an alternative when uncertainty is present in the form of intervals? Selecting among intervals becomes a key in the case of overlapping performance intervals. Given this situation, Decision Theory[10,15] can be implemented, with an emphasis on decisions under strict uncertainty (no likelihood or probability information available). For selection under strict uncertainty, 4 main criteria have been developed: the Maximin (ranks alternatives based on the maximization of the minimum bound), Maximax (rank alternatives based on the maximization of the maximum bound), Hurwicz (rank alternatives based on a weighting of both the minimum and maximum bound), and the Laplace (ranks alternatives based on a equal likelihood of both the minimum and maximum bound) criterion.

The only criterion the authors consider suitable (in the sense of selecting for RM) is the Hurwicz criterion. The Hurwicz criterion takes into account the decision maker's risk preferences, while additionally allowing him/her to grade their risk preference. The Hurwicz criterion also takes into account all the information available, without supplying additional, unfounded, information. When the Hurwicz criterion is used, the decision is considered neither completely pessimistic, nor optimistic. Hurwicz[16] suggests that few would wish to be as pessimistic or optimistic as the Maximin and Maximax criterion suggest. He proposes a optimism-pessimism index, α, where $0 \le \alpha \le 1$, as a measure of the decision-maker's risk preferences. In this criterion, the coefficient of optimism is multiplied by the maximum bound and the coefficient of pessimism, $1-\alpha$, is multiplied by the minimum bound. The critical step in the application of this criterion is the determination of α. To determine α, we employ a method used in utility theory to determine a decision-maker's preference. Specifically, the decision-maker is asked to determine his/her certainty equivalent for a specific lottery, which is a hypothetical situation used to assess the decision-maker's preference[17]. The certainty equivalent is the achievement value at which a designer would be indifferent between receiving the achievement value *for certain* or receiving the result of the lottery between either getting 0 (lowest level of achievement) and 1 (highest level of achievement). The decision maker must choose $\upsilon_{certain}$ such that he/she is indifferent to Option A and Option B in Fig. 11-1.

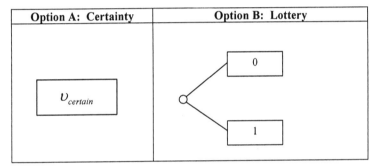

Figure 11-1. Certainty Equivalent Determination

Once the value of $\upsilon_{certain}$ is determined, the value α can be determined using Eq. (1).

$$\alpha = (1 - \upsilon_{certain}) \tag{1}$$

For $0.5 < \alpha \leq 1$, the decision-maker is considered optimistic, where $\alpha = 1$ being completely optimistic (maximax criterion). For $0 \leq \alpha < 0.5$, the decision-maker is considered pessimistic, where $\alpha = 0$ being completely pessimistic (maximin criterion).
The selection criterion for Hurwicz criterion is shown in Fig. 11-2.

> **Selection Criterion**
> Consider the choice between alternative A_i and jth alternatives A_j, where j=1,...,n
> α = optimism-pessimism index, where $0 \leq \alpha \leq 1$
> $$\beta_i = \alpha \cdot MF(A_i)_{max} + (1-\alpha) \cdot MF(A_i)_{min}$$
> and
> $$\beta_j = \alpha \cdot MF(A_j)_{max} + (1-\alpha) \cdot MF(A_j)_{min}$$
> One should select alternative A_i if, and only if:
> $$\beta_i > \beta_j$$

Figure 11-2. Hurwicz Selection Criteria

As can be seen in Figure 11-2, the alternative is selected based upon a weighted sum of the minimum and maximum performance bounds for the alternatives. The weight, α, is derived from the decision maker's risk preferences. The Hurwicz selection parameter, β, is then used to rank and select alternatives.

In this section, we have introduced the concepts of uncertainty and selection under uncertainty, and how we will deal with them in our selection method. In the next section, we will introduce our method for Selection for Rapid Manufacturing under Epistemic Uncertainty. This method builds upon the Selection Decision Support Problem (DSP), developed in[18].

In the Selection DSP, the alternatives are ranked from most feasible to least feasible based on the calculated merit function (MF) value[19]. The higher the merit function, the higher the preference in selection. "Successful applications of DSPs include design of ships, damage tolerant structural and mechanical systems, the design of aircraft, mechanisms, thermal energy systems, design using composite materials and data compression"[19]. Due to the uncertainty involved with the RM selection process, the Selection DSP will be augmented in section 3 to account for the uncertainties inherent to RM.

11.3 SELECTION FOR RAPID MANUFACTURING

In this section, a detailed description of the selection method proposed in this paper is presented. The word formulation for selection for RM is presented in Fig. 11-3.

Selection for Rapid Manufacturing under Epistemic Uncertainty Word Formulation

Given: A set of feasible alternatives.

Identify: The principle *attributes* influencing selection and relative importance of the attributes.

Epistemic uncertainty associated with the geometric dimensions of the part.

The decision maker's risk preference

Assess: Geometric Characteristics that affect the attributes

Rate: The alternatives with respect to each attribute.

Rank: The feasible alternatives *using the Hurwicz criterion* in order of preference based on the attributes and their relative importances

---

Figure 11-3. The Word Formulation for Selection for Rapid Manufacturing

A summary of the steps involved in its implementation are presented in Fig. 11-4.

Steps of Selection for Rapid Manufacturing (RM) under Epistemic Uncertainty

1. Characterize the uncertainty involved
   a. Qualitatively define the Range of Customization
   b. Quantitatively define the uncertainty involved
2. Describe alternatives and provide acronyms
3. Describe each relevant attribute, specify its relative importance, and provide acronyms
4. Specify levels and/or intervals for each attribute for each alternative
5. Normalize the attribute ratings
6. Rank and select the alternatives in order of preference
   a. Evaluate the merit functions
   b. Determine decision maker's risk preferences
   c. Evaluate selection values and rank alternatives
7. Post Solution Analysis and Verification of results

Figure 11-4. Summary of Steps for Selection for Rapid Manufacturing

A detailed description of the 7 steps displayed in Fig. 11-4 is presented next.

Step 1. Characterize the uncertainty involved
Step 1a. Qualitatively define the Range of Customization

Although this step should be done earlier in the design process, it is good practice to re-evaluate the range of customization that is to be offered with a particular product. First, the decision maker should determine the level of customization that is desired. Da Silveira et al. (2001) define eight generic levels of customization, which include: (8) Design (products developed according to individual customer needs), (7) Fabrication (manufacturing of customer-tailored products following basic, predefined designs), (6) Assembly (arranging modular components), (5) Additional custom work, (4) Additional services, (3) Package and distribution, (2) Usage, and (1) Standardization.

Once the level of customization has been determined, the range of customization can be defined qualitatively. In determining the range of customization, the designer should evaluate and describe which features will be customized for the user. For example, a custom footwear designer (Level 8-Design) may qualitatively define a range of customization as customized insoles and standardized uppers and outsoles.

Step 1b. Quantitatively define the uncertainty involved

After the range of customization has been defined qualitatively, a quantitative assessment must be performed. The size ranges for the features can now be defined quantitatively. The designer should define the geometric dimensions the will be constrained (certain) and which will be customized (uncertain).

As stated before, interval analysis will be used to represent the epistemic uncertainty due to customization. When using the interval representation of epistemic uncertainty, uncertainty is modeled with a closed interval bounded by z_{min} and z_{max} (i.e., $Z \in [z_{min}, z_{max}]$). When epistemic uncertainty is modeled using interval numbers, the design equations are converted to intervals[11].

Step 2. Describe the alternatives and provide acronyms

Describe each alternative in words, including its advantages and disadvantages, and provide meaningful acronyms for each. If possible, provide illustrations of the alternatives.

Step 3. Describe each relevant attribute, specify its relative importance and provide acronyms

The next step in solving the selection DSP is to identify the attributes (criteria) by which the alternatives will be evaluated. Depending on the demands of each problem, the attributes will vary. All relevant attributes should be included. The description of each attribute should be comprehensive and understandable. Also, provide meaningful acronyms for the attributes.

In order to specify the relative importance of the attributes, a pair-wise comparison is used. It is noted that other methods, such as the ranking method, can be used. In the pair-wise comparison method, each of the attributes is rated as better than, worse than, or equal to each of the other attributes. For the comparison, a value of 1 given to the attribute that is better, where a 0 is given to the other attribute. If the attributes are considered equal, both attributes receive a value of zero. Next, the values for each attribute are summed and normalized to ensure the relative importance sum to one. To prevent an attribute receiving a total value of zero, a dummy attribute is introduced. In this comparison, the attribute is always preferred to the dummy.

Step 4. Specify scales, rate the alternatives with respect to each attribute.

There are four main types of scales: ratio, interval, ordinal, and composite. The type of information available determines the type of scale chosen. The ratio scale is used when quantitative, physically meaningful units are available for an attribute. When an attribute can only be qualified in words, use the ordinal scale. The interval scale is used to convert the words from an ordinal scale to numerical intervals. The composite scale is used when the value of attribute is the result of computations, such as relative importance analysis.

In this step, the alternatives are also rated with respect to each attribute. The bounded geometric characteristics (such as part volume, area, etc.) are calculated using interval arithmetic operations on the bounded and constrained geometric dimensions. The bounded geometric characteristics will be used to determine selection attributes, such as build time, part cost, etc., which will also be intervals once the uncertainty is propagated. The remaining attributes are also rated using scalar values.

Step 5: Normalize the attribute ratings

Once the alternatives are rated with respect to the attributes, the ratings must be normalized to find a common ground for comparison. When higher values of an attribute rating are preferred, Eq. (2) should be used to normalize the attribute ratings[20]:

$$NR_{ij} = \frac{A_{ij} - A_{j,\min}}{A_{j,\max} - A_{j,\min}} \quad (2)$$

When lower values are preferred, Eq. (3) should be used to normalize the attribute ratings[20]:

$$NR_{ij} = \frac{A_{j,\max} - A_{ij}}{A_{j,\max} - A_{j,\min}} \quad (3)$$

where A_{ij} is the attribute rating w.r.t alternative j, $A_{j,\max}$ is the maximum value of attribute i, and $A_{j,\min}$ is the minimum value of attribute i, and NR_{ij} is the normalized rating of the attribute i w.r.t alternative j.
In the case of uncertainty in the attribute value, NR_{ij} will be bounded as $[NR_{ij}^{\min}, NR_{ij}^{\max}]$.

Step 6: Rank and select the alternatives in order of preference
Step 6a. Evaluate the merit functions
The merit function values of the alternatives are calculated using the Eqs. 4 and 5:

$$MF_{j,\min} = \sum_{j=1} I_i \cdot NR_{ij}^{\min} \quad (4)$$

and

$$MF_{j,\max} = \sum_{j=1} I_i \cdot NR_{ij}^{\max} \quad (5)$$

where MF_j is the merit function of alternative j, I_i is the relative importance of attribute i, and NR_{ij} is the normalized rating of the attribute i w.r.t alternative j. It should be noted that the normalized ratings are relative, therefore, the merit function values' minimum and maximum bounds may be equal.

Step 6b. Determine decision maker's risk preferences
The Hurwicz criterion was chosen because of its accounting of the decision maker's preferences and its inclusion of all the available information, without assuming additional information about the likelihood of an event to occur. The selection criteria for the Hurwicz criterion is displayed in Fig. 11-2.
The decision maker's risk preference is determined by choosing $\upsilon_{certain}$ such that he/she is indifferent to Option A and Option B in Fig. 11-1. Once the value of $\upsilon_{certain}$ is determined, the value α is determined using Eq. (1).

Step 6c. Evaluate selection values and rank alternatives
Instead of being evaluated based on the merit function values, the alternatives are evaluated based on the value, β. β is calculated using Eq. (6).

$$\beta = \alpha \cdot MF_{max} + (1-\alpha) \cdot MF_{min} \tag{6}$$

Once β is calculated, the alternatives are ranked and the top alternative selected based on this value. Higher values of β establish preference.

Step 7: Post Solution Analysis and Verification of results
In this step, the results from Step 6 are reviewed and verified. The designer must determine if the results seem logical and reasonable. Verification may involve changing the weighting schemes (relative importances) of the attributes for different scenarios. Once the merit functions are recalculated, the alternative rankings should be compared and evaluated.

11.4 ILLUSTRATIVE EXAMPLE: DIRECT PRODUCTION OF CASTER WHEELS

In this section, we present the example of the design of caster wheels for Rapid Manufacturing. Using the selection for RM technologies, we will attempt to select a metal RM technology for direct production of steel caster wheels. In this scenario, the caster wheel manufacturer is attempting to select a RM machine that can be used for production of its small custom orders. It is infeasible to stock all the combinations of wheels that they offer, thus there is a need to be able to produce these quickly, while also keeping the price down for the customer. In this situation, RM is believed to be a great candidate. The technologies under consideration are Direct Metal Deposition, Direct Metal Laser Sintering, Electron Beam Melting, Laser Engineered Net Shaping, Selective Laser Melting, and Selective Laser Sintering. A model of a caster wheel is displayed in the Fig. 11-5a.

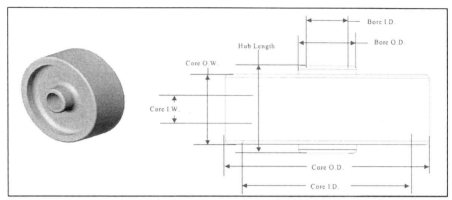

Figure (11-5a). Model of steel caster wheel *(11-5b)* Caster wheel side profile

Before beginning the selection process, the uncertainty involved in the customization process was considered. Since these caster wheels will be customized, there is a degree of geometric uncertainty involved.

Step 1. Characterize the uncertainty involved
In this step, the range of customization is qualitatively assessed. In this example, we have decided to only allow customization of certain features. This example will only deal with the customization of all-steel caster wheels. It should also be noted that only standard 1.25 in. diameter x 4 in. length bolts will be used for the inner bore, therefore these dimensions will be constrained. The customers will be allowed to customize all other features of the caster wheel.

After the range of customization is defined qualitatively, a quantitative assessment must be performed. The designer should define which geometric dimensions will be constrained (certain) and which will be bounded (uncertain). The profile of the caster wheel is displayed in Fig. 11-5b.

The uncertainty is quantified using constraints and bounds on the above dimensions. The constraints and bounds used for this example are displayed in Table 11-1.

Table 11-1. Caster wheel dimensions

	Dimensions	
	min	max
Core Outer Diameter	4	6
Core Inner Diameter	3.5	5.5
Bore Outer Diameter	1.5	2.25
Bore Inner Diameter	1.25	1.25
Hub Length	2.5	2.5
Core Outer Width	1.5	3
Core Inner Width	0.5	1.25

As displayed in the table, the uncertain dimensions are displayed as interval sets. The constrained dimensions are constrained by the standard size of the bolt used in the assembly process.

Step 2. Describe the alternatives and provide acronyms
The alternatives are as follows: Direct Metal Laser Sintering (DMLS), Direct Metal Deposition (DMD), Electron Beam Melting (EBM), Laser Engineered Net Shaping (LENS), Selective Laser Sintering (SLS), and Selective Laser Melting (SLM). Descriptions have been omitted for brevity.

Step 3. Describe each relevant attribute, specify its relative importance and provide acronyms
The attributes are described as follows:

> Ultimate Tensile Strength (UTS): UTS is the maximum stress reached before a material fractures.
> Rockwell Hardness C (Hard): Hardness is the commonly defined as the resistance of a material to indentation.
> Density (Dens.): The density refers to the final density of the part after all processing steps. This density is proportional to the amount of voids found at the surface. These voids cause a rough surface finish.
> Detail Capability (DC): The detail capability is the smallest feature size the technology can make.
> Geometric Complexity (GC): The geometric complexity is the ability of the technology to build complex parts. More specifically, in this case, it is used to refer to the ability to produce overhangs.
> Build Time (Time): The build time refers to the build time of a part, not including post processing steps.
> Part Cost (Cost): The part cost is the cost it takes to build one part with all costs included. These costs include manufacturing cost, material cost, machine cost, operation cost, etc.

In this example, we examine 2 weighting scenarios (relative importance ratings). In Scenario 1, a pairwise comparison was used to determine relative importance of each attribute. In this scenario, geometric complexity was most heavily weighted because of the significant overhangs present in the build orientation of the casters. Build time and part cost were also heavily weighted because of their importance to the business structure surrounding customization of caster wheels. Because of the environment of use of the caster wheels, UTS was also given a high weighting. Detail capability was weighted least because of the lack of small, detailed features in the geometry of the caster wheels. In Scenario 2, all selection attributes were equally weighted. The relative importance weightings for each scenario are presented in Table 11-3.

Step 4: Specify scales, rate the alternatives with respect to each attribute.

At this step, bounded geometric characteristics (such as part volume, area, etc.) are calculated using interval arithmetic operations on the bounded and constrained geometric dimensions. In our case, the particular the geometric constraint of concern are the bounded part volume, which is used to calculate the build time and part cost. These will be used to determine the build time and part cost of the parts. The bounded part volume is [168923.6 mm^3, 1170399 mm^3].

The alternative ratings are presented in Table 11-2.

Table 11-2. Attribute Ratings

						Attributes					
		UTS	Hard	Density	Detail Cap.	Geom. Compl.		Build Time_avg		Part Cost	
						Min	Max	Min	Max	Min	Max
Rel. Imp	Scen. 1	0.167	0.143	0.071	0.024	0.214	0.214	0.190	0.190	0.190	0.190
	Scen. 2	0.143	0.143	0.143	0.143	0.143	0.143	0.143	0.143	0.143	0.143
Alternatives	DMD	1800	53	100	1.016	4	6	0.12	1.85	29.48	168.15
	DMLS	600	21	95	0.3	7	10	2.83	58.90	386.98	2045.93
	EBM	1430	50	100	1.2	7	10	1.42	29.56	134.41	508.56
	LENS	1703	53	100	0.762	4	6	0.26	4.76	64.17	306.52
	SLM	2000	60	99.5	0.15	7	10	1.88	38.98	237.43	1340.57
	SLS	606	15	100	0.6	7	10	1.42	23.56	180.67	889.63
Scales	Type	Ratio	Ratio	Ratio	Ratio	Interval		Ratio		Ratio	
	Pref.	high	high	high	low	high		low		low	
	Units	Mpa	HRc	percent	mm	nmu		hrs		USD	

Step 5: Normalize the attribute ratings
The attribute ratings in Table 11-2 were normalized using the equations presented in Section 3.

Step 6: Evaluate the merit functions
The merit function values of the alternatives (Scenario 1 and 2) are displayed in Table 11-3. After performing the lottery in Fig. 11-1, the author determined $\upsilon_{certain}$=0.3, where α was computed to be 0.7 from Eq. 1. Using α calculated above and the Merit Function Values from Table 11-3, the Hurwicz evaluation parameter, β, was calculated using Eq. 6 for both scenarios. These values are found in Table 11-4.

After performing the lottery in Fig. 11-1, the author determined $\upsilon_{certain}$=0.3, where α was computed to be 0.7 from Eq. 1. Using α calculated above and the Merit Function Values from Table 11-3, the Hurwicz evaluation parameter, β, was calculated using Eq. 6 for both scenarios. These values are found in Table 11-4.

Table 11- 3. Alternative Merit Function Values for Scenario 1 and 2

	Merit Function			
	Scenario 1		Scenario 2	
	min	max	min	max
DMD	0.720	0.720	0.697	0.697
DMLS	0.254	0.254	0.284	0.284
EBM	0.729	0.750	0.657	0.672
LENS	0.686	0.691	0.700	0.704
SLM	0.759	0.750	0.804	0.810
SLS	0.509	0.535	0.525	0.544

Table 11- 4. Hurwicz evaluation parameters

	Scenario 1	Scenario 2
DMD	0.720	0.697
DMLS	0.254	0.284
EBM	0.743	0.667
LENS	0.689	0.703
SLM	0.753	0.808
SLS	0.528	0.539

Step 7: Post Solution Analysis and Verification of results
As seen in Table 11-4, for Scenario 1, SLM and EBM came out atop the other alternatives. This is largely due to the fact that a heavy importance weighting was given to geometric complexity, as well as build time and build cost. This scenario favors SLM and EBM since they both use powder beds (which favor production of overhangs), as well as significantly greater volumetric build rates (quoted) than the remaining alternatives. In Scenario 2, equal importance was given to all attributes. Although SLM, LENS, and DMD are all ranked relatively close to one another, SLM was a clear top performer. It can also be seen that in Scenario 1, LENS was ranked 4, as opposed to a top rank in Scenario 2 (and vice versa for EBM). This is due to the fact that with equal attribute weightings, geometric complexity becomes less of a factor. It should be noted that similar results can be expected from all parts within the volumetric range determined in Step 4 (all else equal). Based on our knowledge of the metal RM processes, the rankings seem in order, given the conditions specified in the example.

11.5 ILLUSTRATIVE EXAMPLE: DIRECT PRODUCTION OF HEARING AID SHELLS

Using the selection for RM technologies, we will attempt to select a RM technology for direct production of custom hearing aid shells in this section. Given the nature of the hearing aid business and competition, there is a need to be able to produce these quickly and cheaply, while also mimicking the quality exhibited by hand-manufactured products. Given this need, most hearing aid companies already use RM to produce custom hearing aid shells. A rendering of a typical hearing aid shell is displayed in Figure 11-6.

Figure 11-6. Hearing Aid Shell[21]

Due to customization, each hearing aid will be different in a manner which is difficult to quantify parametrically. Because of this, we have chosen to represent the hearing aid as an elliptical cone, with the following parameters: major diameter, minor diameter, height, and wall thickness. Since these hearing aid shells are custom, there is a considerable degree of uncertainty in each of the above parameters.

Step 1. Characterize the uncertainty involved
In this step, the range of customization is qualitatively defined. For this example, we decided to allow full customization of all the dimensions of the hearing aid shell, except the wall thickness, which is fixed at 1.1 mm. The uncertainty is quantified using constraints and bounds on the dimensions of the hearing aid shells, displayed in Table 11-5.

Table 11-5. Hearing Aid Shell Dimensions

	Dimensions	
	min	max
major diameter	13	18
minor diameter	8	11
height	16	22
thickness	1.1	1.1

Step 2. Describe the alternatives and provide acronyms
In this example, the alternatives are combinations of RM machines and materials. We chose 3 different RM technology groups: 3D Systems' Stereolithography (SLA 5000, SLA 7000, and SLA viper systems), 3D Systems' Selective Laser Sintering (Sinterstation HiQ system), and Stratasys' Fused Deposition Modelling (FDM Titan system). For the stereolithography (SLA) systems, Renshape SL5510, Renshape SL7560, DSM Somos 10120, and DSM Somos 9120 resins were used. For the selective laser sintering (SLS) systems, Duraform PA and Duraform GF powders were used. For the fused deposition modeling (FDM) system, ABS P400 was used.

Step 3. Describe each relevant attribute, specify its relative importance

and provide acronyms

The attributes are described as follows:
Ultimate Tensile Strength (UTS): (see Section 4).
Young's Modulus (YM): YM is used to indicate the stiffness of the material.
Flexural Strength (FS): FS is the measure of a material's ability to resist bending.
Flexural Modulus (FM): FM is used to indicate the bending stiffness of the material.
Build Time (Time): (see Section 4).
Part Cost (Cost): (see Section 4).

In this example, we examine 2 weighting scenarios (relative importance ratings). In Scenario 1, a pairwise comparison was used to determine relative importance of each attribute. In this scenario, build time and part cost were most heavily weighted because of their importance to the business structure surrounding customization of hearing aid shells. Flexural modulus was also highly weighted because of its direct impact on the customer. The relative importance weightings for each scenario are presented in Table 11-6.

Step 4. Specify scales, rate the alternatives with respect to each attribute.

At this step, bounded geometric characteristics (such as part volume, area, etc.) are calculated using interval arithmetic operations on the bounded and constrained geometric dimensions. In our case, the particular the geometric constraint of concern are the bounded part volume, which is used to calculate the build time and part cost in a build time estimation software package. The bounded part volume is [115.3 mm^3, 224.9 mm^3]. The alternative ratings are presented in Table 11-6.

Step 5: Normalize the attribute ratings
The attribute ratings in Table 11-6 were normalized using the equations presented in Section 3.

Step 6: Rank and select the alternatives in order of preference
The merit function values for Scenario 1 and Scenario 2 are displayed in Table 11-7. As in Example 1, α was computed to be 0.7 from Eq. (1). Given α and the merit function values, the Hurwicz evaluation parameter, β, was calculated using Eq. 6 for both scenarios. These values are also found in Table 11-7.

Table 11-6. Attribute Ratings

		Attributes							
		Tensile Strength	Young's Mod	Flex. Str.	Flex. Mod	Build Time_avg		Part Cost	
						Min	Max	Min	Max
Rel. Imp	Scen. 1	0.086	0.086	0.171	0.200	0.229	0.229	0.229	0.229
	Scen. 2	0.167	0.167	0.167	0.167	0.167	0.167	0.167	0.167
Alternatives	SLA5000.9120	31	1344.5	43.5	1382.5	0.0039	0.0082	0.42	0.88
	SLA5000.10120	26	1710	39.5	1310	0.0039	0.0082	0.42	0.88
	SLA5000.7560	52	2500	93.5	2500	0.0039	0.0082	0.42	0.88
	SLA5000.5510	77	3296	99	3296	0.0039	0.0082	0.42	0.88
	SLA7000.9120	31	1344.5	43.5	1382.5	0.0022	0.0049	0.30	0.64
	SLA7000.10120	26	1710	39.5	1310	0.0022	0.0049	0.30	0.64
	SLA7000.7560	52	2500	93.5	2500	0.0022	0.0049	0.30	0.64
	SLA7000.5510	77	3296	99	3296	0.0022	0.0049	0.30	0.64
	SLAviper.9120	31	1344.5	43.5	1382.5	0.0072	0.0162	0.71	1.54
	SLAviper.10120	26	1710	39.5	1310	0.0072	0.0162	0.71	1.54
	SLAviper.7560	52	2500	93.5	2500	0.0072	0.0162	0.71	1.54
	SLAviper.5510	77	3296	99	3296	0.0072	0.0162	0.71	1.54
	SLS_Duraf_PA	44	1600	44	1285	0.0033	0.0063	0.33	0.64
	SLS_Duraf_GF	38	5910	38.1	3300	0.0033	0.0063	0.33	0.64
	FDM_Tit_ABS	35	2480	34.5	2495	0.0145	0.0288	1.29	2.55
Scales	Type	Ratio	Ratio	Ratio	Ratio	Ratio		Ratio	
	Pref.	high	high	high	high	low		low	
	Units	MPa	MPa	MPa	MPa	hrs/ part		USD	

Rapid Prototyping: Theory and Practice 289

Table 11-7. Merit Function Values and Hurwitz factors

	Scenario 1			Scenario 2		
	M.F. min	M.F. max	Hurwicz Factor	M.F. min	M.F. max	Hurwicz Factor
SLA5000.9120	0.441	0.439	**0.44**	0.338	0.337	**0.34**
SLA5000.10120	0.421	0.420	**0.42**	0.319	0.318	**0.32**
SLA5000.7560	0.741	0.740	**0.74**	0.671	0.670	**0.67**
SLA5000.5510	0.892	0.890	**0.89**	0.862	0.860	**0.86**
SLA7000.9120	0.499	0.500	**0.50**	0.381	0.381	**0.38**
SLA7000.10120	0.480	0.480	**0.48**	0.362	0.362	**0.36**
SLA7000.7560	0.800	0.800	**0.80**	0.713	0.714	**0.71**
SLA7000.5510	0.951	0.951	**0.95**	0.904	0.905	**0.90**
SLAviper.9120	0.313	0.284	**0.29**	0.245	0.224	**0.23**
SLAviper.10120	0.294	0.265	**0.27**	0.226	0.205	**0.21**
SLAviper.7560	0.614	0.585	**0.59**	0.578	0.557	**0.56**
SLAviper.5510	0.765	0.735	**0.74**	0.769	0.747	**0.75**
SLS_Duraf_PA	0.492	0.505	**0.50**	0.407	0.417	**0.41**
SLS_Duraf_GF	0.747	0.760	**0.76**	0.696	0.706	**0.70**
FDM_Tit_ABS	0.157	0.157	**0.16**	0.171	0.171	**0.17**

Step 7: Post Solution Analysis and Verification of results

As seen in Table 11-7, for Scenario 1 and 2, SLA7000 using 5510 resin ranked atop the other alternatives, followed by SLA5000 using 5510 resin. This is mainly due to the superior material properties of the 5510 resin, as well as the high build speed and low part cost of the SLA 5000 and 7000 machines. In our example, the stereolithography machines seem to outperform the other technologies in most cases. Based on our knowledge of the RM processes, the rankings seem in order, given the conditions specified in the example.

11.6 CLOSURE

So what do these results mean? In this selection process, the uncertainty is introduced in Step 4 in the form of a minimum and maximum geometric bounds on the dimensions of the part being produced (caster wheels in first case hearing aid shells in second case). This uncertainty was set by the amount of customization that was offered by the designer. With that, this uncertainty is propagated (using interval arithmetic) to get a bounded part volume, which was used to determine the alternative ratings for build time and part cost. Once the ratings are calculated, the uncertainty is propagated to the merit function values. The alternatives are then ranked based on the Hurwicz selection parameter, which takes into account the risk preference of the decision-maker, as well as the uncertainty in the merit function values. To the working engineer, this method gives a way to account for the uncertainty that is inherent to customization in the selection process. This method allows the decision maker to account for the wide array of customized parts that can be built using RM in a single selection process. Otherwise, a separate evaluation process for each geometry being produced in the RM machine is needed.

In this paper, we presented a systematic method for selection of a Rapid Manufacturing technology under the epistemic uncertainty inherent to customization. The method was demonstrated using examples of RM selection for the production of caster wheels and custom hearing aid shells.

ACKNOWLEDGEMENTS

This work has been supported by the David and Lucille Packard Foundation Fellowship, Harriett G. Jenkins Predoctoral Fellowship Program, and by the industry members of the Georgia Tech Rapid Prototyping & Manufacturing Institute. The assistance of various members of the Systems Realizations Laboratory at Georgia Tech is gratefully acknowledged.

References

1. T. Wohlers, Wohlers Report 2004: Rapid Prototyping, Tooling, & Manufacturing State of the Industry (2004).
2. M.G. Fernandez, C.C. Seepersad, D.W. Rosen, J.K. Allen, and F. Mistree, Utility-Based Decision Support for Selection in Engineering Design. ASME Design Engineering Technical Conference and Computers and Information in Engineering Conference. 2001. Pittsburgh, Pennsylvania: ASME (2001).
3. L. Cheng, E. Subrahmanian, and A.W. Westerberg, Design and planning under uncertainty: issues on problem formulation and solution. Computers & Chemical Engineering, **27,** 781-801(2001).
4. J.N. Siddall ed, Probabilistic Engineering Design., Marcel Dekker: New York (1983).
5. S. Vadde, J.K. Allen, and F. Mistree, Catalog Design: Selection using available assets, Engineering Optimization, **25,** 45-64 (1995).
6. J.K Allen, The Decision to Introduce New Technology: The Fuzzy Preliminary Selection Decision Support Problem. Engineering Optimization, **26,** 61-77 (1996).

7. R.P. Reddy, and F. Mistree, Modeling Uncertainty in Selection using Exact Interval Arithmetic, Design Theory and Methodology, ASME.(1992).
8. H.-R. Bae, R.V. Grandhi, and R.A. Canfield, Epistemic uncertainty quantification techniques including evidence theory for large-scale structures, Computers & Structures, **82**, 1101-1112 (2002).
9. P.S. Laplace, Essai Philosphique sur les Probabilites. 1825, Paris: Translation published by Dover, New York (1952).
10. S. French, Decision Theory: An Introduction to the Mathematics of Rationality, Chichester: Ellis Horwood Limited (1986).
11. S.S. Rao, and L. Cao, Optimum Design of Mechanical Systems Involving Interval Parameters, ASME Journal of Mechanical Engineering, **124**, 465-472 (2002).
12. R.E. Moore, Interval Analysis, Englewood Cliffs, NJ: Prentice-Hall (1996).
13. H.M. Regan, M. Colyvan, and M.A. Burgman, A Taxonomy and Treatment of Uncertainty for Ecology and Conservation Biology. Ecological Applications, **12**(2), 618-628 (2002).
14. S. Ferson, and L.R. Ginzburg, Different methods are needed to propagate ignorance and variability. Reliability Engineering and System Safety, **54**, 133-144 (1996).
15. A. Rapoport, Decision Theory and Decision Behavior: Normative and Descriptive Approaches, Dordrecht/ Boston/ London: Kluwer Academic Publishers (1989).
16. L. Hurwicz, Optimality criteria for decision making under ignorance, in Cowles Commision Discussion Paper No. 370 (mimeographed) (1951).
17. M.G. Fernandez, On Decision Support for Distributed Collaborative Design and Manufacture, Mechanical Engineering, Georgia Institute of Technology, Atlanta, GA (2002).
18. F. Mistree, W.F. Smith, B. Bras, J.K. Allen, and D. Muster. Decision-Based Design: A Contemporary Paradigm for Ship Design, Transactions, Society of Naval Architects and Marine Engineers, Jersey City, New Jersey (1990).
19. A. Hermann, and J.K. Allen, Selection of Rapid Tooling Materials and Processes in a Distributed Design Environment, ASME Design Engineering Techical Conferences, Las Vegas, Nevad, SME (1999).
20. E. Bascaran, R.B. Bannerot, and F. Mistree, Hierarchical Selection Decision Support Problems in Conceptual Design, Engineering Optimization, **14**, 207-238 (1989).
21. M. Masters, Direct Manufacturing of Custom-Made Hearing Instruments, SME Rapid Prototyping Conference, Cincinnati, OH (2002).

Chapter 12

ECONOMIC ANALYSIS OF RAPID PROTOTYPING SYSTEMS

Khaled M. Gad El Mola[1], Hamid R. Parsaei[2], and Herman R. Leep[3]

[1]*Visiting Professor, Department of Industrial Engineering, University of Houston, Houston, TX 77204;* [2]*Professor and Chairman, Department of Industrial Engineering, University of Houston, Houston, TX 77204;* [3]*Professor, Department of Industrial Engineering, University of Louisville, Louisville, KY 40292*

Abstract

Considering the large sums of money required to implement advanced manufacturing technologies (AMT), management must carry out appropriate evaluations and then make critical decisions. Traditional justification methods are insufficient by themselves because they cannot cope with benefits, such as flexibility and enhanced quality, offered by AMT. Selection and justification processes for AMT involve complex problems and require extensive analysis of a large number of criteria. An appropriate decision-making procedure for justification of AMT requires consideration of both economic and non-economic investments. An analytical model is presented for the selection of a rapid prototyping (RP) system from a set of mutually exclusive decision alternatives. The proposed model is based on the analytic hierarchy process (AHP) along with the Expert Choice software and it provides the means for integrating economic with non-economic benefits. A numerical example illustrates an application of the proposed model.

Keywords:
Advanced manufacturing technologies (AMT), rapid prototyping systems, economic analysis, analytic hierarchy process (AHP), Expert Choice.

12.1. Introduction

In order to survive in today's global competitive environment, manufacturing companies must be capable of manufacturing high-quality products at low cost with increasing variety over short lead times and then be able to deliver them to customers at the right time. In order to achieve these objectives and to respond to these challenges, many companies have adopted advanced manufacturing technologies (AMT) such as computer-aided design (CAD), manufacturing resource planning (MRP II), flexible manufacturing systems (FMS), computer-integrated manufacturing (CIM), rapid prototyping technology, and so forth. These technologies have been used extensively in industry. AMT can be considered as a means for increasing productivity, quality, flexibility, and profitability in order to cope with the pressure of international competition. Although AMT can improve overall manufacturing performance, companies maintain some ambivalence toward deciding whether to invest in such technologies. Also, the subject of the economics of AMT recently has become of significant interest to the management science community[42].

Selection and justification of AMT involve decisions that are critical to the profitability and survival of a manufacturing company in an increasingly competitive environment. AMT requires a high capital outlay yet offers a large number of non-economic benefits such as flexibility, competitiveness, customer satisfaction, and so on which are generally difficult to quantify. Traditional justification approaches are insufficient to justify AMT. Researchers have adopted numerous approaches to justify AMT, but these approaches often provide limited information. For example, economic justification approaches are incapable of analyzing intangible attributes. Justification of AMT is an unstructured problem, and thus it requires extensive analysis of a large number of variables.

The major hurdles in implementing AMT are not engineering shortcomings in the equipment or manufacturing processes, but rather, managerial attitudes and policies are examples in the justification problem[25].

Today, most major corporations are struggling with their traditional justification methods because these methods are either misapplied, or the information included in the calculations is inadequate for the multifaceted problem being tackled[6]. Manufacturing technologies reshape the basic ways in which a firm perceives the problems, and hence, they often alter the firm's objectives. Decision-making in such situations becomes quite complex because manufacturing excellence encompasses quality, flexibility, throughput times, customer satisfaction, and many other such problems. Basically, the justification problem for manufacturing systems is a very difficult one to solve because of the following[6]:

- Cost patterns of new technologies are rarely understood, so cost estimates and pricing policies for innovative products are fraught with uncertainty.
- Traditional methods of accounting do not consider non-economic benefits such as quality, reduced lead times, and so on.
- Investment in islands of technology within the company does not add any significant improvements.
- Companies frequently set extremely high goals for investment opportunities with the notion that high discount rates ensure high returns. Often, just the opposite holds true if a long-term perspective is taken.
- The trap of sunk costs tends to bind future technology decisions to previous decisions.
- Capital investment is high.
- Payback period is quite substantial, hence, it is difficult to apply traditional approaches.
- Cost patterns have changed, manufacturing overheads continue to grow for many companies, and costs such as maintenance, quality control and material handling should be done on an individual basis.
- Usability of the technology is not being considered.

Rapid prototyping is a relatively new technology. Wozny[51] defines it as the process of selectively depositing a gas, liquid, powder, or sheet material in layers. The purpose of these processes is to produce solid three-dimensional parts directly from CAD models[32].

Thus, the objective of this chapter is to analyze the problem of justifying advanced manufacturing technologies by using a multiple criteria decision-making (MCDM) approach such as analytic hierarchy process (AHP) and the Expert Choice software.

The next section in this chapter provides a comprehensive review of recent literature on justification techniques for AMT. A brief presentation of the concepts of AHP and the Expert Choice software will be presented in the third section. An analytical model to justify AMT is presented in the fourth section and illustrated with an example in the fifth section. Finally, conclusions are drawn in the final section of the chapter.

12.2. Recent Literature on Justification Techniques

Researchers have contributed much toward justifying automation technologies using various justification approaches, and they have also categorized these approaches. For example, Meredith and Suresh[24] partitioned the literature on the justification methods of advanced manufacturing technologies into three groups that correspond to different levels of integration of new manufacturing technology – from stand-alone systems to completely integrated systems such as CIM. Their first group, called "economic justification approaches," is concerned with simple economic functions such as payback (PB), return on investment (ROI), internal rate of return (IRR), net present value (NPV), and so on, used in problems where there is no uncertainty.

Investments in equipment such as CNC machine tools are often replacement projects, and economic justification approaches are appropriate for such projects[24]. These approaches have been adopted by researchers for the justification problem. For example, Rygh[36] applies a payback method for the justification of an automated storage and retrieval system. Klahorst[21] used ROI, NPV, and PB for the justification of flexible manufacturing systems. Fotsch[13] conducted a mail survey of 67 manufacturing companies and found that 65% of these firms named the payback method, while 26% named ROI as the most frequently used justification methods. Primrose and Leonard[34] reported that in the United Kingdom, 41% of the firms used IRR, 32% used PB, 32% used ARR, and only 17% used NPV as the technique for investment appraisal. Miltenburg and Krinsky (1987) analyzed the application of traditional economic justification approaches to the evaluation of the FMS alternatives. Canada and Sullivan[6] explained the benefit/cost analysis for project selection using economic factors.

Ease of data collection, simplicity, and clarity are the primary advantages of the traditional economic justification approaches; however, the inability of these variables to incorporate strategic benefits that may not be quantifiable in financial terms stands out as their major drawback. Although major strategic benefits such as early entry into the market, market leadership, innovation, improved flexibility, and quality are of vital importance in the justification of advanced manufacturing technologies, they are not readily expressed in cash flow terms.

The second group is called "analytic justification approaches," and it includes those models which are too complex for the first group and too quantitative for the third group. When flexibility, risk, and non-monetary benefits are expected, as with rapid prototyping, analytical justification approaches are required. In order to justify manufacturing systems at the intermediate level of integration (namely, CAD, CAM, FMS, and so on), analytic approaches such as risk analysis, portfolio analysis, and value analysis have been suggested[24].

Risk analysis is an approach that conducts simulation studies to examine important variables such as costs, benefits, yield, and capacity, and to expresses the results statistically and/or graphically[47]. Various automation projects can thus be simulated beforehand and the results compared. Pike and Ho[33] observed that the reason for the lack of widespread acceptance of risk analysis methods by firms in assessing capital investment projects is due to the fact that they are marred by inherent and practical problems still to be resolved. They listed difficulties and problems in the use of risk analysis. They also conducted a comprehensive survey in the United Kingdom, which covered a wide range of risk analysis issues in capital budgeting.

Portfolio analysis includes three categories of techniques, namely, non-numeric, scoring, and mathematical programming. Non-numeric techniques appear to be non-scientific in nature, although their use in the real world cannot be ruled out. Meredith and Suresh[24] provided two different models, sacred cow and operating necessity, as non-numeric techniques.

Scoring models[8] seem to have rational underlying principles, and their use has been reported by several authors[31]. The weighted factor scoring model developed by Satty[37] is called the "Analytic hierarchy process" (AHP). Sprinivasan and Millen[44] used AHP as the basis of a framework to evaluate the strategic impact of FMS investment, and they have attempted to integrate such evaluations with traditional discounted cash flow (DCS) procedures. Arbel and Seidmann[3] used AHP to evaluate and compare two similar systems, such as two FMSs. Wabalickis[49] developed a justification methodology based on AHP to evaluate the many tangible and intangible benefits of an FMS investment. Mohanty and Venkataraman[29] used the AHP for justification of three automated manufacturing systems.

Mathematical programming (integer, goal, and quadratic) is the third type of portfolio analysis technique. These models have been developed to quantify performance measures such as quality and flexibility for justifying investment in advanced manufacturing systems. Nelson[31] presented a scoring model for FMS project selection, which supplements traditional capital budgeting procedures with the treatment of non-economic criteria and project interdependency.

Value analysis in this context is an adaptation of the value analysis technique suggested by Keen[20] for justifying decision support systems. It is a prototyping technique that builds a system in stages. Hwang and Yoon[16] developed the Technique for Order Preference by Similarity to Ideal Solution (TOPSIS) based upon the concept that the chosen alternative should have the shortest distance from the ideal solution and the farthest from the negative solution. Agrawal et al.[1] used TOPSIS for selection of a robot. Agrawal et al.[2] also adapted TOPSIS for the selection and evaluation of optimum grippers for a robot.

The third group from Meredith and Suresh[24] is called "strategic justification approaches." These approaches are qualitative in nature, and are concerned with issues such as technical importance, business objectives, competitive advantage, and research and development. When strategic justification approaches are employed, the justification is made by considering long-term intangible benefits. The problem with these approaches is that they are extremely subjective, and they lack adequate structures to be used with sufficient confidence as valid scientific methods.

Naik and Chakravarty[30] reviewed various justification approaches. They summarized the justification techniques according to four main types: economic, analytical, strategic, and integrated approaches. Most integrated approaches are developed for projects with different combinations of economic, analytic, and strategic approaches. Sloggy[43] suggested a multi-attribute utility model, a somewhat more formal way of combining financial and non-financial factors in justifying new technology systems. De and Whinston[7] developed a framework for complex manufacturing problems. The methodology is a decision support system (DSS) framework that focuses on key decisions and tasks, with the specific aim of improving the effectiveness of the manager's problem solving process. Garrett[15] suggested a combination of strategic and financial methods, but he used a three-level approach to screen proposals. Sprinivasan and Millen[44] developed a framework to integrate explicitly a systemic assessment of the strategic impact of FMS investment with the traditional discounted cash flow techniques. They attempted to integrate the evaluations of AHP with traditional discounted cash flow (DCF) procedures to evaluate the strategic impact of the FMS investment.

Stam and Kuula[45] developed a two-phase decision procedure that uses AHP and multi-objective mathematical programming to select a FMS. Suresh[46] proposed a DSS structure for flexible automation investments. The model takes into account the investment context for a multi-machine system, economic evaluation, and an integrated physical/financial design and evaluation. Madu and Georgantzas[23] adopted a DSS framework for the strategic selection of computer integrated manufacturing. They used a group decision making strategy for conflict resolution and the AHP for idea analysis. Mladineo et al.[27] developed a DSS for multi-criteria analysis in public policy decision. Naik and Chakaravarty[30] developed a three-level framework which combined strategic, tactical, and operational aspects for strategic acquisition of AMT. Ramasesh and Jayakumar[35] proposed a four-stage justification framework using traditional DCF and mathematical programming.

Shang and Sueyoshi[41] proposed a selection procedure for a FMS employing the AHP, simulation, and data envelopment analysis. Sambasivarao and Deshmukh[39] presented a DSS integrating multi-attribute analysis, economic analysis, and risk evaluation analysis. They suggested AHP, TOPSIS, and the linear additive utility model as alternative multi-attribute analysis methods.

Mohanty[28] classified various justification approaches into four classes: qualitative, semi-qualitative, quantitative, and mathematical programming models. Qualitative models generally deal with the identification of problems and the causes of such problems rather than finding a specific solution. Such methods are basically insightful and facilitate the creation of a perturbation in the mind of the decision maker by exposing his/her decisions to a variety of factors and attributes. Semi-qualitative methods are used to transform the subjective judgments to simple linear measures in order to evaluate the AMT. Quantitative methods aim at quantifying the varieties of attributes of AMT basically through monetary or surrogate measures. Mathematical models seek to determine the best (optimum) course of action for a decision problem under the restriction of limited resources, when the objectives and the constraints of the model can be expressed quantitatively or mathematically as functions of the decision variables. Thus a mathematical model is formed. Table 12-1 presents a summary of some of the justification approaches documented in the literature.

It is apparent from the literature that the justification of AMT is a complex process and involves a large number of attributes. The selection of an AMT is problem-specific and is an unstructured problem.

12.3. Analytic Hierarchy Process and Expert Choice

Analytic Hierarchy Process (AHP) is a decision-aiding tool for dealing with complex, unstructured, and multi-criteria decisions. It was developed at the Wharton School of Business by Thomas Saaty in the late 1970s. It is a flexible model that allows individuals or groups to shape ideas and solve problems by making their own assumptions. The AHP incorporates judgments and personal values in a logical way – the same way the human mind sorts through multiple options in the course of decision-making. The decision-making process depends on experience, imagination, and knowledge to structure a problem and upon experience, intuition, and logic to provide judgments for solving the problem[38]. Unlike conventional decision-analysis techniques, the AHP methodology does not require numerical guessing. It can easily accommodate subjective judgments concerning aspects of a problem for which no scale of measurement exists. These judgments are used in deriving ratio scale priorities for the decision criteria and alternatives[9]. The theory of AHP is based on the three stages of problem solving. They are the principles of decomposition, comparative judgments, and synthesis of priorities.

Table 12-1. Summary of Available Justification Approaches

Economic Approaches	<ul><li>Payback</li><li>Net present value</li><li>Internal rate of return</li><li>Return on investment</li><li>Non-DCF</li><li>Sensitivity analysis</li></ul>	
Analytic Approaches	**Risk Analysis** <ul><li>Stochastic methods</li><li>Monte Carlo simulation</li></ul>	
	Portfolio Analysis	**Non-numeric**<ul><li>Sacred cow</li><li>Operating necessity</li></ul>
		Scoring<ul><li>AHP</li><li>Unweighted 0-1</li></ul>
		Programming<ul><li>Integer</li><li>Goal</li><li>Quadratic</li></ul>
	Value Analysis<ul><li>Utility models</li></ul>	
Strategic Approaches	<ul><li>Technical benefits</li><li>Business advantage</li><li>Competitive factors</li><li>Research and development</li></ul>	
Integrated Approaches	<ul><li>Multi-attribute utility theory</li><li>Scenario planning and multilevel screening</li><li>Computer programming</li><li>Simulation and DCF</li><li>AHP and DCF</li><li>Expert systems</li><li>Manufacturing system value</li><li>AHP and mathematical programming</li><li>AHP, simulation, and DEA</li><li>DSS framework</li></ul>	

A sensitivity analysis can be performed to see how sensitive the alternatives are to changes in the importance of the criteria, players, or environmental scenarios. There are five types of sensitivity analysis available with Expert Choice software for the implementation of AHP, according to Dyer and Forman[10] and Forman et al.[12]. They include gradient sensitivity, dynamic sensitivity, performance sensitivity, 2-dimensional plotting, and head-to-head graphing.

Expert Choice (EC) is the microcomputer software implementation of AHP. Developed by Forman et al.[11], EC has led to a growing number of applications of AHP in a variety of decision-making problems. Expert Choice accommodates pairwise comparisons verbally, numerically, and graphically[10]. The newer version of Expert Choice (Advanced Decision Support Software) includes all the features of the earlier versions but has a new feature called Expert Choice Group (ECG). ECG is a decision-support software designed to help groups enhance the quality of their decisions by bringing structure to the decision-making process. It enables group members any place in the world to make decisions together through the Internet. More details about the features of EC may be found in the Expert Choice Group software manual[12].

Multi-Criteria-Decision-Making (MCDM) methods provide a comprehensive set of qualitative and quantitative criteria to help in justifying the AMT. In fact, subjective measurement is the only concept that is widely accepted in MCDM to deal with multi-criteria problems. A robust MCDM procedure used for justification of AMT should be able to incorporate tangible and intangible factors. AHP is one of the MCDM methods that which has been used very successfully in many situations in which a decision situation is characterized by a multitude of complementary and conflicting factors. Therefore, AHP has been selected as the simplest tool that provides accurate models in the context of a performance measurement system. Many researchers[18, 40, 5, 4, 48] have compared AHP with five other approaches, including the utility approach and multiple regression approach. The results showed that AHP is the least difficult and most robust approach, as well as being highly useful in performance measurement.

12.4. Analytical Model to Justify Advanced Manufacturing Technologies

A proposed analytical model to justify advanced manufacturing technologies using the standard AHP and Expert Choice will be presented in this chapter. This model combines the economic and non-economic benefits and consists of the following main steps:
1. Identify the overall objective of the analysis, which is to select the best alternatives from advanced manufacturing technologies.

2. List the set of AMT decision alternatives (A_i, $1 \leq i \leq m$) that the organization can undertake. Three AMT decision alternatives will be considered in this analysis. They are as follows: rapid prototyping, software prototyping (CAD/CAM), and hardware prototyping.
3. Identify the competitive criteria and their measures, which may have an impact on the organization.
4. Structure the hierarchy.
5. Determine the overall weight for each alternative and select the alternative which has the highest weight.

12.4.1. Identify the competitive criteria and their measures

Competitive criteria and identification of their sub-criteria is the most crucial step in this model. Failing to include them in the model may result in deterioration of the outcomes. A review of the existing literature on manufacturing systems was performed in order to identify the competitive criteria by focusing on the competitive strategy. The manufacturing competitive criteria (quality, flexibility, productivity, time, cost, customer satisfaction, market share, finance, and delivery) appear to be the most widely accepted in the manufacturing strategy literature[22, 19, 50, 17]. Those critical criteria can be seen to cover all aspects of a manufacturing system: financial results, operating performance (through the dimensions of quality, flexibility, time, delivery and cost), and the way the company is perceived externally (through its customers). Gad El Mola[14] identifies more than 200 measures as being relevant to the following competitive criteria: quality, flexibility, time, delivery, cost, customer satisfaction, and finance.

12.4.2. Structuring the hierarchy

Once competitive priorities and their measures have been identified, the decision-maker begins to structure the hierarchy. This model proposes a hierarchy with four management levels. They include the top, upper middle, lower middle, and operational levels. In order to recognize the ever-increasing importance of non-economic benefits, at the top management level, the hierarchy consists of two criteria: non-economic and economic categories. At one level lower than the top management level (the first-middle management level), the measures include manufacturing system criteria such as quality, flexibility, dependability, customer satisfaction, time, and productivity, and economic criteria such as profitability, market

Rapid Prototyping: Theory and Practice 303

share, and cost. In addition, the criteria at this level may have several sub-criteria out of which is constructed the second-middle management level. For example, the quality criterion has two sub-criteria: customer response and conformance to specifications. The time criterion has two cub-criteria: internal and external. The dependability criterion has two sub-criteria: delivery and machine availability. The productivity criterion has three sub-criteria: partial, total-factor, and total productivity. The lowest management level (operational management level) includes measures that lead to simplification of the manufacturing process compared to high-level performance measures. The last level includes the AMT decision alternatives, which can contribute to each operational measure in a unique way. The hierarchical scheme just explained is shown in Figure 12-1.

12.4.3. Determine the overall weight for each alternative and select the alternative that has the highest weight

Once the hierarchy has been structured, the overall weight of each alternative can be calculated. The following steps describe in detail how to determine the overall weight for each alternative.

1. Assign weights to each category at the top management level on the basis of the relative importance of its contribution to the overall objective. Thus, one input pairwise comparison of criteria matrix $PM_t = (\alpha_{tij})$ of n_t by n_t will be obtained. Here, n_t is the total number of categories at the top management level, and the element α_{tij} is the importance of criterion i compared to criterion j at this level. A weighting procedure within the AHP enters PM_t and generates an eigenvector $RW_t = (rw_{ti})$ that has n_t elements. Typically, $n_t = 2$ (economic and non-economic benefits).

2. At the second level of the hierarchy, consider the ith category of the top management level. Then, one input matrix of pairwise comparisons of criteria (at the first-middle level) that corresponds to the ith criterion of the top level can be obtained. The result is stored in an n_{ml} by n_{ml} matrix PM_{ml}^i. Here, n_{ml} is the total number of criteria at this level. A weighting routine in enters puts PM_{ml}^i and generates the relative weight for each criterion in relation to the ith criterion at one level higher. This "local relative weight" is stored in a local weighting vector RW_{ml}^i. Typically, in comparison with "global" weights, "local" weights are obtained within the framework of single criterion at one level higher. The "global relative weight" at this level is computed and then stored in a vector $GW_{ml} = (gw_{mli})$ that has the n_{ml} elements as follows:

304 Rapid Prototyping: Theory and Practice

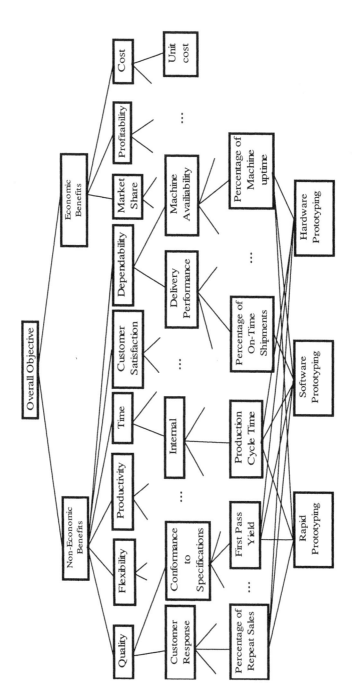

Figure 12-1. Hierarchical Structure for Justification of Advanced Manufacturing Technologies

$$GW_{m1} = RW_t \times \begin{pmatrix} RW_{m1}^1 \\ \cdot \\ \cdot \\ \cdot \\ RW_{m_1}^{n_1} \end{pmatrix} \qquad (1)$$

3. The second step will be repeated for the second middle level and for the operational level, and the global relative weights will be obtained as GW_{m2} and $GW_o = (gw_{oj})$.
4. Assign weights to each alternative on the basis of the relative weight of its contribution to each operational performance measure. Thus, one input pairwise comparison of alternative matrix PM_L^j of n_L by n_L will be obtained. Here, n_L is the total number of decision alternatives at the Lth level. A weighting procedure within the AHP enters PM_L^j and generates an eigenvector $RW_L^j = (rw_{Lj})$ that has n_L elements.
5. The overall weight for each alternative (OW_{A_i}) can be calculated using the following equation, and then the alternative with the highest weight can be selected.

$$OW_{A_i} = \sum_{i=1}^{n_L}\sum_{j=1}^{n_o} GW_o * RW_L^j \qquad (2)$$

12.5. An Illustrated Example

For simplicity, the variables flexibility, time, and customer satisfaction will be selected as non-economic criteria. For economic criteria, profitability and cost will be selected. Also, the sub-criteria will not be considered in this example. The operational measures are product change response, design/training time, cycle time, delivery reliability, start-up costs, tooling costs, and return on investment (ROI). Figure 12-2 shows the AHP model for selecting the best AMT alternative. First, the relative importance of non-economic and economic criteria are stored in a vector: $RW_t = (0.4, 0.6)$.

Figure 12-3 shows a graphical pairwise comparison of economic and non-economic benefits with respect to the overall objective. Next, the management level is considered. The relative weights of the criteria in this level with relation to each top level criterion are to be computed. First, the local relative weights are calculated. For the non-economic and economic criteria, the pairwise comparison matrices are PM_m^1 and PM_m^2, respectively. The relative weights will be stored in the vectors RW_m^1 and RW_m^2, respectively.

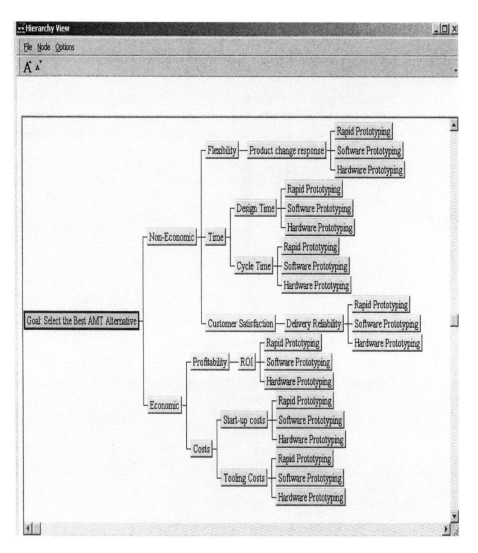

Figure 12-2. AHP Model for Selecting the Best AMT Alternative

$$\text{PM}_m^1 = \begin{bmatrix} 1 & 4 & 2 \\ 1/4 & 1 & 1/3 \\ 1/2 & 3 & 1 \end{bmatrix} \qquad \text{PM}_m^2 = \begin{bmatrix} 1 & 1/8 \\ 8 & 1 \end{bmatrix}$$

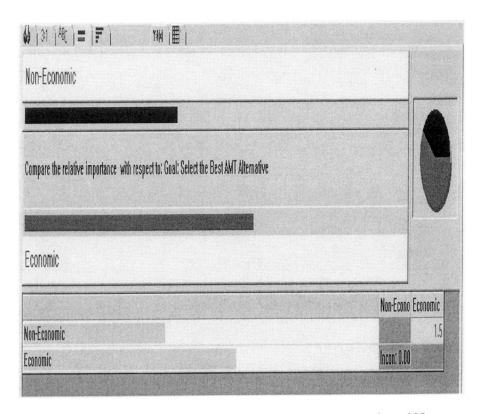

Figure 12-3. A Graphical Pairwise Comparison of Economic and Non-Economic Benefits with Respect to the Overall Objective

Figure 12-4 shows a numerical pairwise comparison of non-economic criterion. Here, for the non-economic criteria, flexibility is estimated to be four times more important than time and twice as important as customer satisfaction. Customer satisfaction is estimated to be three times more important than time. As a result, $\text{RW}_m^1 = (0.558, 0.122, 0.32)$.

For each weight computation, an inconsistency ratio (γ) was computed and checked for acceptance; i.e., in this case, the result is accepted because $\gamma = 0.0174 \leq 0.10$. Figure 12-5 shows the relative weights with respect to non-economic benefits. Figure 12-6 shows the verbal pairwise comparisons of economic criterion. For the economic criterion, profitability was estimated to be eight times more important than cost. As a result, $RW_m^2 = (0.111, 0.889)$.

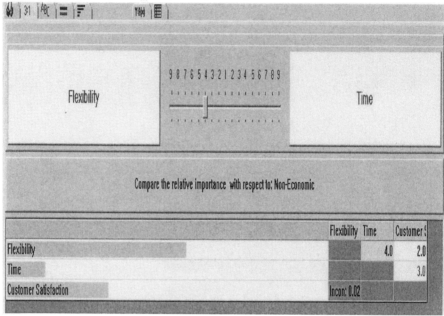

Figure 12-4. A Numerical Pairwise Comparison of Non-Economic Criterion

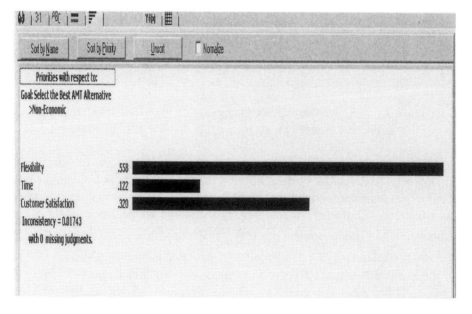

Figure 12-5. Relative Weights with Respect to Non-Economic Benefits

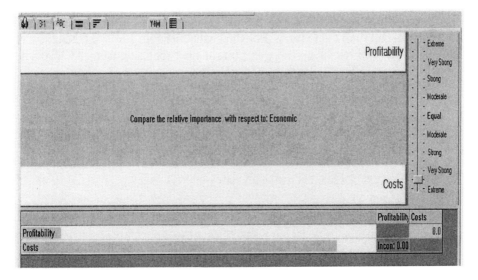

Figure 12-6. A Verbal Pairwise Comparison of Economic Criterion

Accordingly, the global relative weights of the managerial criteria (in order of flexibility, time, customer satisfaction, profitability, and cost) are as follows:

$$GW_m = (0.4, 0.6) \times \begin{pmatrix} 0.588, 0.122, 0.32, 0, 0 \\ 0, 0, 0.111, 0.889 \end{pmatrix} = (0.2232, 0.0488, 0.1280, 0.0666, 0.5334)$$

Here, the global relative weights of flexibility, time, customer satisfaction, profitability, and cost are 22.32%, 4.88%, 12.80%, 6.66%, and 53.34%, respectively. Note that these percentages are elements of the above GW_m.

Now, move down to the operational level. The relative weights of the operational measures in this level with relation to each top level criterion are to be computed. For the time and costs criteria, the pairwise comparison matrices are PM_o^2 and PM_o^5, respectively. The relative weights are stored in the vectors RW_o^2 and RW_o^5, respectively:

$$PM_o^2 = \begin{bmatrix} 1 & 1/4 \\ 4 & 1 \end{bmatrix} \qquad PM_o^5 = \begin{bmatrix} 1 & 1/3 \\ 3 & 1 \end{bmatrix}$$

Here, the time criterion and design time is estimated to be four times more important than cycle time. For the costs criterion, start-up costs time is estimated to be three times more important than tooling costs. As a result, $RW_o^2 = (0.20, 0.80)$ and $RW_o^5 = (0.75, 0.25)$.

Accordingly, the global relative weights of the operational measures (in order of product change response, design time, cycle time, delivery reliability, ROI, start-up costs, and tooling costs) are as follows:

$$GW_o = (0.2232, 0.0488, 0.128, 0.0666, 0.5334) \times$$

$$\begin{pmatrix} 1,0,0,0,0,0,0 \\ 0,0.2,0.8,0,0,0,0 \\ 0,0,0,1,0,0,0 \\ 0,0,0,0,1,0,0 \\ 0,0,0,0,0,0.75,0.25 \end{pmatrix} = (0.2232, 0.00976, 0.03904, 0.128, 0.0666, 0.40005, 0.13335)$$

Finally, the pairwise comparisons of the three AMT alternatives with respect to each operational measure is stored in the matrices PM_L^j, and the relative weight will be stored in the vectors RW_L^j. For product change response operational measure,

$$PM_L^1 = \begin{bmatrix} 1 & 2 & 5 \\ 1/2 & 1 & 2 \\ 1/5 & 1/2 & 1 \end{bmatrix} \quad RW_L^1 = (0.595, 0.276, 0.128),$$

Figure 12-7 shows the relative weights of the three alternatives with respect to product change response measure. Similarly, for other operational measures one level higher, the local weights are computed as:

$RW_L^2 = (0.667, 0.222, 0.111) \quad RW_L^3 = (0.705, 0.211, 0.084)$

$RW_L^4 = (0.550, 0.240, 0.210) \quad RW_L^5 = (0.682, 0.216, 0.103)$

$RW_L^6 = (0.727, 0.2, 0.073) \quad\quad\quad RW_L^7 = (0.571, 0.286, 0.143)$.

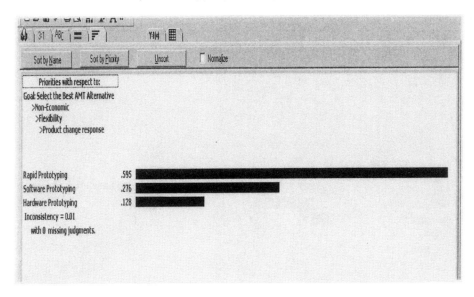

Figure 12-7. Relative Weights of Alternative with Respect to Product Change Response

Figures 12-8 to 12-13 show the relative weights of the three alternatives with respect to product change for each operational measure.

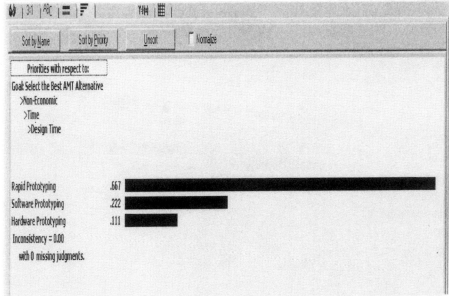

Figure 12-8. Relative Weights of Alternatives with Respect to Design Time

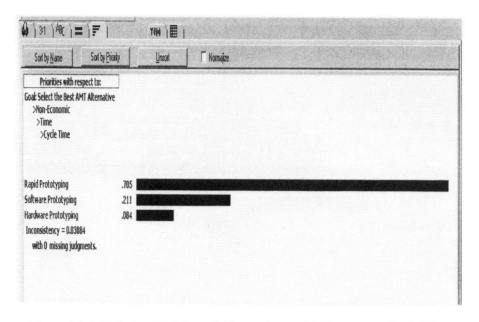

Figure 12-9. Relative Weights of Alternatives with Respect to Cycle Time

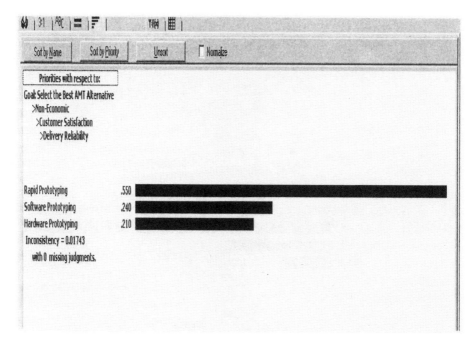

Figure 12-10. Relative Weights of Alternatives with Respect to Delivery Reliability

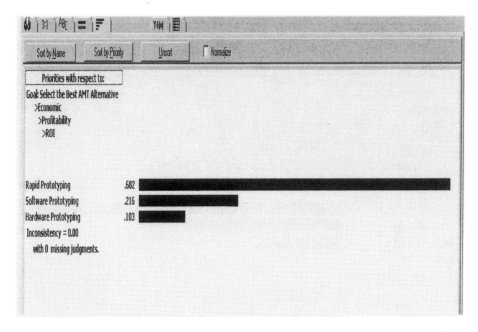

Figure 12-11. Relative Weights of Alternatives with Respect to ROI

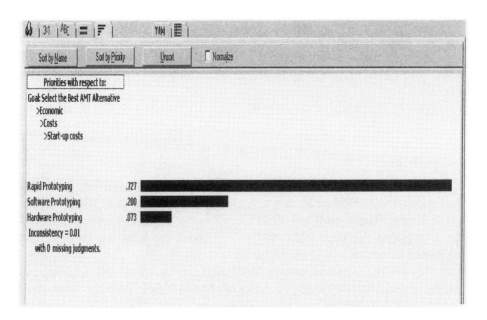

Figure 12-12. Relative Weights of Alternatives with Respect to Start-Up Costs

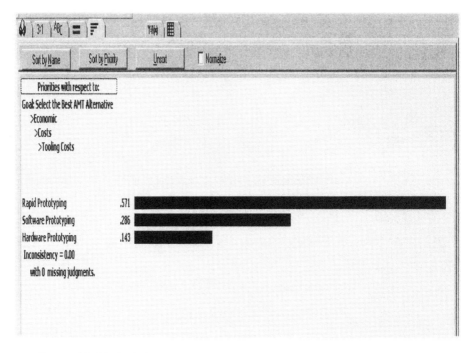

Figure 12-13. Relative Weights of Alternatives with respect to Tooling Costs

Applying Eq. (2) in this example, the overall weight for the first alternative (rapid prototyping) is calculated as follows:

$$OW_{A_1} = 0.2232 \times 0.595 + 0.00976 \times 0.667 + 0.03904 \times 0.705 + 0.1280 \times 0.550$$
$$+ 0.0666 \times 0.682 + 0.40005 \times 0.727 + 0.13335 \times 0.571 = 0.650$$

Similarly, the overall weights for the other two alternatives (software and hardware prototyping) are calculated. They are 0.235 and 0.115, respectively. The overall weight of the RP is the highest (0.650) and should be selected as the best AMT alternative given the benefits, criteria, and measures examined in this example.

The above analytical model is dynamic over time. As a company evolves, the hierarchy must be adjusted accordingly. Another interesting aspect is that the hierarchy is not company-invariant. The hierarchy must be adjusted depending on the unique situation faced by each individual company or division. Basically, the hierarchical schema is designed to accommodate any number of levels. New levels can easily be added to or deleted from the hierarchy. Therefore, the proposed model is highly flexible. It incorporates economic and non-economic measures into the hierarchy. Furthermore, these two types of measures are compared indirectly; i.e., an economical measure is compared with another economical measure, while a non-economical measure is compared with another non-economical measure, respectively. This comparison method relieves the burdens of human estimators because of homogeneous comparisons.

AHP is a flexible tool because it can be applied to any hierarchy of justification problems, regardless of the number of levels of competitive criteria and their measures. However, there are some assumptions that have to be carefully considered before applying AHP to the problem of justification. Such assumptions, if disregarded, can mislead managers about this technique. The assumptions for the proposed model are as follows:

1. Competitive priorities and their measures should be independent, not redundant and additive;
2. The pairwise comparisons made by managers have to be fairly consistent;
3. The 1 to 9 preference scale allows the relative importance of competitive priorities and their measures to be expressed well.

12.6. Summary

This chapter presented an overview of available methods for the justification of advanced manufacturing technologies. Uncertainty concerning market conditions forces manufacturing companies to look for more and more advanced technologies. Large investments involve critical decisions and require an appropriate justification analysis. In recent times, the adoption of economic and multi-attribute decision approaches are called upon to aid in breaking down, analyzing, communicating, and synthesizing the nature of the justification problem, and hopefully lead to the best decision under the circumstances. Selection of an AMT alternative is thus a complex problem. The proposed analytical model integrates economic and non-economic benefits to justify AMT alternatives. It is flexible and dynamic over time. A numerical example was included to illustrate the steps of the model and also to show how the Expert Choice software is beneficial when applied to the AHP.

References

1. V. P. Agrawal, V. Kohli, and S. Gupta, Computer aided robot selection: the multiple attribute decision making approach, International J. Production Research, **29**, 1629-1644 (1991).

2. V. P. Agrawal, A. Verma, and S. Agrawal, Computer aided evaluation and selection of optimum grippers, International J. Production Research, **30**, 2713-2732 (1992).
3. A. Arbel and A. Seidmann, Performance evaluation of FMS, IIE Transactions on Systems, Man, and Cybernetics, **14**, 606-617 (1984).
4. J. F. Bard, A comparison of the analytic hierarchy process with multiattribute utility: the case study, IIE Transactions, **24**, 111-121 (1992).
5. V. Belton, A comparison of the analytic hierarchy process and a simple multi-attribute function, European J. of Operational Research, **26**, 7-21 (1986).
6. J. R. Canada and W. Sullivan, Economic and Multi-attributes Evaluation of Advanced Manufacturing Systems (Prentice Hall, Englewood Cliffs, NJ, 1989).
7. S. De and A. B. Whinston, , A framework for integrated problem solving in manufacturing, IIE Transactions on Industrial Engineering, **18**, 286-297 (1986).
8. B. V. Dean, Evaluating, Selecting and Controlling R&D Projects (American Management Association, New York, 1968).
9. R. F. Dyer, and E. H. Forman, An Analytic Approach to Marketing Decisions (Prentice Hall, Englewood Cliffs, NJ. 1991).
10. R. F. Dyer and E. H. Forman, Group decision support with the analytic hierarchy process, Decision Support Systems, **8**, 99-124 (1992).
11. E. H. Forman, T. L. Saaty, M. A. Selly, and R. Waldron, Decision Support Software Expert Choice Inc. (Mclean, VA, 1983).
12. E. H. Forman, T. L. Saaty, A. Shvartsmsan, M. R. Forman, M. Korpics, J. Zottola, and M. A.Selly, , Expert Choice: Advanced Decision Support Software, (Expert Choice, Inc., Pittsburgh PA, 2002)
13. R. J. Fotsch, Machine tool justification policies: their effect on productivity and profitability, J Manufacturing Systems, **2**, 169-195 (1983).
14. K. M. Gad El Mola, A methodology to measure the performance of manufacturing systems, Ph.D. Dissertation, University of Houston, Houston, TX (2004).
15. S. E. Garett, Strategy first: a case in FMS justification. Proceedings of 2nd ORSA/TIMS Conference on FMS: OR models and applications, Edited by K. E. Steck, and R Suri, Elsevier Science Publishers, Amsterdam, pp. 17-29 (1986).
16. C. L. Hwang and K. S. Yoon, Multiple Attribute Decision Making: A State-of-the-Art Survet (Springer-Verlag, New York, 1982).
17. M. Hudson, A. Smart, and M. Bourne, Theory and practice in SME performance measurement systems, International J. Operations & Production Management, **21**, 1096-1115 (2001).

18. R. D. Kamenentzky, The relationship between the AHP and the additive value fur, Decision Science, **13**, 702-713 (1982).
19. R. S. Kaplan and D. P. Norton, The balanced scorecard – measures that drive performance, Harvard Business Review, **70**, 71-79 (1992).
20. P. G. Keen, Value Analysis: Justifying Decision Support Systems, MIS Quarterly, March 5 (1981).
21. H. T. Klahorst, How to Justify Multi-Machine System, in: Machinist, edited by J. R. Meredith, 1983.
22. G. K. Leong, D. L.Snyder, and P. T. Ward, Research in the process and content of manufacturing strategy, OMEGA International Journal of Management Science, **18**, 109-122 (1990).
23. C. N. Madu and N. C.Georgantzas, Strategic thrust of manufacturing automation decisions: a conceptual framework, IEE Transactions on Industrial Engineering, **23**, 138-148 (1991).
24. J. R. Meredith and N. C. Suresh, Justification techniques for advanced manufacturing technologies, International J. Operations Research, **24**, 1043-1057 (1986).
25. G. J. Michael and R. A. Millen, Economic justification of modern computer-based factory automation equipment, Annals Operations Research, **3**, 25-34 (1985).
26. G. J. Miltenburg and I. Krinsky, Evaluating flexible manufacturing systems, IIE Transactions, **19**, 222-233 (1987).
27. N. Mladineo, I. Lozie, S. Stosie, D. Mlinare, and T. Radiea, An evaluation of multicriteria analysis for DSS in public policy decision, European J Operational Research, **61**, 219-229 (1992).
28. R. P. Mohanty, Analysis of justification problems in CIMS: review and projections, International J Production Planning and Control, **4**, 260-271 (1993).
29. R. P. Mohanty and S. Venkatraman, Use of analytic hierarchy process for selecting automated manufacturing systems, International J Operations and Production Management,. **13**, 45-57 (1993).
30. B. Naik and A. K. Chakravarty, Strategic acquisition of new manufacturing technology: a review and research framework, International J Production Research, **30**, 1575-1601 (1992).
31. C. A. Nelson, A scoring model for flexible manufacturing systems project selection, European J Operational Research, **24**, 346-359 (1986).
32. M. Philbin, Rapid Prototyping: A Young Technology Evolves, Modern Casting 1996.
33. R. H. Pike and S. M. M. Ho, Risk analysis in capital budgeting: barriers and benefits, Omega, International J Management Science, **19**, 235-245 (1991).
34. P. L. Primrose and R. Leonard, The development and application of a computer based appraisal technique for the structural evaluation of machine tool purchases, Proceedings of Institution of Mechanical Engineers, 198B, 141-146 (1984).

35. R. V. Ramasesh and M. D. Jayakumar, Economic justification of advanced manufacturing technology, OMEGA, International J Management Science, **21**, 289-306 (1993).
36. O. B. Rygh, Justifying an automated storage and retrieval system, Industrial Engineering, **July**, 20-24 (1981).
37. T. L. Saaty, The Analytical Hierarchy Process (McGraw Hill, New York, 1980).
38. T. L. Saaty, Decision Making for Leaders: The Analytical Hierarchy Process for Decisions in a Complex World, (RWS Publications, Pittsburgh, PA, 1990).
39. K. V. Sambasivarao and S. G. Deshmukh, A decision support system for selection and justification of advanced manufacturing technologies, Production Planning and Control, **8**, 270-284 (1997).
40. J. H. Schoemaker, and C. C. Waid, An experimental comparison of different approaches to determining weights in additive utility models, Management Science, **28**, 182-196 (1982).
41. J. Shang and T. Sueyoshi, A unified framework for the selection of a flexible manufacturing system European J Operational Research, **85**, 297-315 (1995).
42. K. Singhal, C. H. Fine, J. R. Meredith, and R. Suri, Research and models for automated manufacturing, Interfaces, **17**, 5-14 (1987).
43. J. J. Sloggy, How to justify the cost of an FMS, Tooling and Production, **50**, 72-75 (1984).
44. V. Sprinivasan and R. A. Millen, Evaluating flexible manufacturing systems as a strategic investment, Proceedings of 2nd ORSA/TIMS Conference on FMS: OR models and application, Edited by K. E. Steck, R. Suri Elsevier Science Publishers, Amsterdam, pp. 83-93 (1986).
45. A. Stam and M. Kuula, Selecting a flexible manufacturing system using multiple criteria analysis, International J Production Research, **29**, 803-820 (1991).
46. N. C. Suresh, Towards an integrated evaluation of flexible automation investments, International J Production Research, **28**, 1657-1672 (1990).
47. E. Turban and J. R. Meredith, Fundamentals of management science (Business Publications, Inc., Plano, Texas, 1985).
48. E. G. Zapatero, C. H. Smith, and H. R. Weistroffer, Evaluating multiple-attribute decision support systems, J Multi-Criteria Decision Analysis, **6**, 201-214 (1997).
49. R. N. Wabalickis, Justification of FMS with analytic hierarchy process, J Manufacturing Systems, **7**, 175-182 (1988).
50. G. P. White, A survey and taxonomy strategy-related performance measures for manufacturing, International Journal of Operations & Production Management, **16**, 42-61 (1996).
51. M. Wozny, Data Driven Solid Freeform Fabrication, Human Aspects in Computer Integrated Manufacturing (North-Holland, New York, 1992).

Index

A
ACR/NEMA, 87,100
Active techniques, 87,90
AFM, 107,127
Analytical Network Process, 253, 254, 255, 260
Assembly analysis model, 134,147, 148, 154, 155
Assembly design formalism, 134, 145, 146, 147
Assembly operation, 133, 134, 135, 136, 138, 139, 145, 146, 147, 148, 151

B
Boundary Representation, 25

C
CAD (computer aided design), 25, 27, 34, 44, 49
CAE, 27, 134, 138, 147
CAM, 27, 87, 89, 100, 110, 111, 116, 117
CCD, 87, 95, 96, 107, 124
Ceramics extrusion, 221
CERR, 79, 81, 82
CNC, 125, 165, 166, 167, 168, 169, 170, 172, 174, 175, 176, 179, 180, 181, 182, 184, 187, 188, 189, 190, 191, 192
CNC Machining, 125, 166, 167, 168, 169, 180, 184, 188, 194
Collaborative product development, 134, 135, 138, 160
Collaborative virtual prototyping and simulation, 134, 163
Computed tomography, 63, 64, 87, 96, 97, 98
Computer Aided Geometric Design (CAGD), 27
Computerized numerical control, 107, 110

Concavity, 39, 50, 51, 52, 55, 58
Concrete extrusion, 221
Conformance, 67, 70, 84
Construction automation, 221, 251
Contact based systems, 107, 108, 120
Contour Crafting, 221, 222, 225, 226, 228, 235, 239, 248, 249, 250
Coordinate measuring machines, 107, 110, 116

D

Data collection, 107, 118
Data Format, 27, 29
Data Transfer, 27, 72, 184
Decision Analysis, 254, 256, 260
Decision Theory, 274
DICOM, 63, 65, 66, 67, 68, 70, 71, 72, 73, 74, 75, 76, 77, 78, 79, 83, 87, 101
DICOM RT, 76, 77
DICOM3, 87
Digitization, 107, 119, 127
Directory Entry Section, 31
Droplet Deposition, 197

E

economic analysis, 88, 293, 298
e-design, 133, 134, 141, 143, 144, 150, 151
e-design brokers, 134, 150, 151
Electronic Health Record, 67
Entity-Relationship Models, 68
Epistemic, 271, 272, 273, 274, 278, 290
e-tools, 134, 137, 143
Expert Choice, 293, 295, 299, 301, 315
Extraterrestrial construction, 221, 248
Extrusion, 138, 157, 221, 222, 223, 227, 228, 229, 233, 242, 243, 244, 247

F

FEA, 134, 144, 151
Feature Recognition, 25, 33, 60, 171, 173, 175, 194
Form Features, 37, 39, 49, 53
Formative process, 104, 221

G
Geometric uncertainty, 271, 273, 274, 282
H
Haptic volume sculpting, 107, 125, 126
House construction, 221
Hurwicz, 272, 274, 275, 280, 284, 288, 290
I
IGES (Initial Graphics Exchange Specifications), 25, 27, 29, 30, 31, 32, 32, 33, 34, 37, 44, 46, 48, 54, 107, 125
Integrating the Healthcare Enterprise, 67
Internet, 65, 133, 135, 137, 138, 141, 143, 144
Interval analysis, 272, 273, 274, 278
Investment Analysis, 254
J
JDICOM, 72
Joining, 93, 134, 135, 136, 138, 139, 145, 151, 198
L
Laplace, 274
large scale fabrication, 221, 237, 239, 249
laser scanning, 87, 91, 92, 93, 101, 108, 169, 213
Layered Fabrication, 197, 249
Layered Manufacturing, 221
Lunar construction, 221
M
Machining, 41, 54, 165, 167, 168, 171, 177, 186, 192
Material Selection, 2, 20, 197, 202, 241, 273
Maximax, 274, 275
Mechanical Properties, 9, 18, 19, 135, 226, 241, 249
Mechanical Testing, 9
Medical image data file, 99
Medical Imaging, 63, 65, 67, 75, 78, 84, 96, 99, 102
Mega scale fabrication, 221
MEMS, 127
MRI (Magnetic Resonance Imaging), 64, 74, 79, 96, 97, 99, 101, 105
MRIcro, 72, 73

Multiattribute Utility Theory, 254
Multiple Criteria Decision Making, 254
N
Nalytical Hierarchy Process, 255, 260
Nano CMM, 127, 128, 129
NC Path Generation, 197
NEMA (National Electrical Manufacturers Association), 64, 65, 68, 71, 100
Non-contact, 89, 95, 102, 104, 105, 108, 120
O
Object Oriented Data Structure, 33, 34, 37, 44
Optimization, 79, 81, 83, 139, 192, 210, 210, 214
P
Parameter Data Section, 31, 32, 34
Passive techniques, 90
Pegasus service manager, 144, 149, 150, 151, 155
Performance parameters, 121
Photogrammetry, 94, 105
Plane Surface (parameterized), 35
Point cloud, 119, 120, 121, 124, 127
Point preprocessing, 107
Polymers, 2, 7, 8, 10, 17, 18, 19, 197, 219
Probe accuracy, 122, 130
Process Planning, 25, 27, 33, 34, 60, 146, 137, 169, 179, 195
R
Radiation Treatment Planning, 78, 86
Rapid Manufacturing, 21, 133, 271, 272, 276, 290
Rapid Prototyping, 1, 7, 18, 83, 134, 159, 196, 213, 226, 260
Reverse engineering, 87, 88, 90, 95, 102, 104, 110, 126
Rigid body errors, 123
Riveting, 139, 148, 151
S
Scanning speed, 108, 110, 121
Selection, 20, 134, 168, 177, 195, 202, 218, 244, 260, 271, 277, 289
Selection Decision Support Problem, 276

Service providers, 141, 144, 149, 151, 156
Service-oriented architecture, 142, 149, 152, 156
Sintering, 19, 166, 197, 202, 219, 254, 281, 287
Sintering Inhibition, 198
SIS, 197, 198, 201, 202, 204, 210, 217, 218, 219
Solid modeling, 25, 26, 27, 257
Standard, 9, 16, 21, 25, 27, 29, 30, 33, 34, 60, 63, 64, 65, 67, 70, 75, 75, 100, 147, 204, 261
STEP THROUGH, 52
STL, 18, 19, 87, 93, 125, 167, 175, 178, 204
STL format, 93, 205
STM, 108, 127
Strategic Technology Justification, 254
strict uncertainty, 274
Structural deformations, 124
Structure of IGES, 30
Surface fitting, 121
Surface of revolution, 35

T

Taxonomy, 87, 89
Technology Management, 254
Thermoplastics extrusion, 221
Thick layer fabrication, 221
Three-dimensional reconstruction, 101

U

Ultrasound scanning, 99
Uncertainty, 124, 136, 255, 272, 273, 278, 286

V

Virtual assembly analysis, 133, 134, 136, 143, 160

W

Welding, 134, 135, 139, 146, 155
Working Group, 65, 67, 83